COMPUTATIONAL METHODS for ELECTRIC POWER SYSTEMS

The ELECTRIC POWER ENGINEERING Series
Series Editor Leo Grigsby

Published Titles

Electromechanical Systems, Electric Machines, and Applied Mechatronics
Sergey E. Lyshevski

Electrical Energy Systems
Mohamed E. El-Hawary

Electric Drives
Ion Boldea and Syed Nasar

Distribution System Modeling and Analysis
William H. Kersting

Linear Synchronous Motors: Transportation and Automation Systems
Jacek Gieras and Jerry Piech

The Induction Machine Handbook
Ion Boldea and Syed Nasar

Power Quality
C. Sankaran

Power System Operations and Electricity Markets
Fred I. Denny and David E. Dismukes

Computational Methods for Electric Power Systems
Mariesa Crow

Forthcoming Titles

Electric Power Distribution Handbook
Tom Short

COMPUTATIONAL METHODS for ELECTRIC POWER SYSTEMS

MARIESA CROW
University of Missouri
Rolla, Missouri

CRC PRESS

Boca Raton   London   New York   Washington, D.C.

Library of Congress Cataloging-in-Publication Data

Crow, Mariesa.
    Computational methods for electric power systems / Mariesa Crow.
        p. cm. — (The electric power engineering series)
    Includes bibliographical references and index.
    ISBN 0-8493-1352-X
    1. Electric power systems—Mathematical models. 2. Electric
engineering—Mathematics. I. Title. II. Series.

TK1005 .C76 2002
333.793′2—dc21
                                             2002035041

To
Jim, David, and Jacob

Preface

This book is the outgrowth of a graduate course that I've taught at the University of Missouri-Rolla for the past decade or so. Over the years, I've used a number of excellent textbooks for this course, but each textbook was always missing some of the topics that I wanted to cover in the class. After relying on handouts for many years, my good friend Leo Grigsby encouraged me to put them down in the form of a book (if arm-twisting can be called encouragement ...). With the support of my graduate students, who I used as testbeds for each chapter, this book gradually came into existence. I hope that those who read this book will find this field as stimulating as I have found it.

In addition to Leo and the fine people at CRC Press, I'm grateful to the University of Missouri-Rolla administration and the Department of Electrical and Computer Engineering for providing the environment to nurture my teaching and research and giving me the latitude to pursue my personal interests in this field.

Lastly, I don't often get the opportunity to publicly acknowledge the people who've been instrumental in my professional development. I'd like to thank: Marija Ilic, who initially put me on the path; Peter Sauer, who encouraged me along the way; Jerry Heydt, for providing inspiration; Frieda Adams, for all she does to make my life easier; Steve Pekarek, for putting up with my grumbling and complaining; and Lowell and Sondra Crow for making it all possible.

Mariesa L. Crow
Rolla, Missouri

Contents

1 Introduction   **1**

2 The Solution of Linear Systems   **3**
2.1 Gaussian Elimination 4
2.2 LU Factorization 9
     2.2.1 LU Factorization with Partial Pivoting 16
     2.2.2 LU Factorization with Complete Pivoting 20
2.3 Condition Numbers and Error Propagation 22
2.4 Relaxation Methods 23
2.5 Conjugate Gradient Methods 28
2.6 Problems 34

3 Systems of Nonlinear Equations   **37**
3.1 Fixed Point Iteration 38
3.2 Newton-Raphson Iteration 42
     3.2.1 Convergence Properties 45
     3.2.2 The Newton-Raphson for Systems of Nonlinear Equations 46
     3.2.3 Modifications to the Newton-Raphson Method 50
3.3 Power System Applications 51
     3.3.1 Power Flow 52
     3.3.2 Regulating Transformers 60
     3.3.3 Decoupled Power Flow 64
     3.3.4 Fast Decoupled Power Flow 66
     3.3.5 PV Curves and Continuation Power Flow 69
     3.3.6 Three-phase Power Flow 76
3.4 Problems 78

4 Sparse Matrix Solution Techniques   **81**
4.1 Storage Methods 82
4.2 Sparse Matrix Representation 87
4.3 Ordering Schemes 89
     4.3.1 Scheme 0 97
     4.3.2 Scheme I 98
     4.3.3 Scheme II 104
     4.3.4 Other Schemes 106
4.4 Power System Applications 108
4.5 Problems 112

5 Numerical Integration   **117**
    5.1   One-Step Methods 118
        5.1.1   Taylor Series-based methods 118
        5.1.2   Forward-Euler method 119
        5.1.3   Runge-Kutta methods 119
    5.2   Multistep Methods 120
        5.2.1   Adam's methods 126
        5.2.2   Gear's methods 129
    5.3   Accuracy and Error Analysis 130
    5.4   Numerical Stability Analysis 134
    5.5   Stiff Systems 141
    5.6   Step-Size Selection 145
    5.7   Differential-Algebraic Systems 148
    5.8   Power System Applications 149
        5.8.1   Transient Stability Analysis 150
    5.9   Mid-Term Stability Analysis 159
    5.10 Problems 162

6 Optimization   **167**
    6.1   Least Squares State Estimation 168
        6.1.1   Weighted Least Squares Estimation 171
        6.1.2   Bad Data Detection 174
        6.1.3   Nonlinear Least Squares State Estimation 177
    6.2   Steepest Descent Algorithm 178
    6.3   Power System Applications 184
        6.3.1   Optimal Power Flow 184
        6.3.2   State Estimation 195
    6.4   Problems 200

7 Eigenvalue Problems   **205**
    7.1   The QR Algorithm 206
        7.1.1   Shifted QR 212
        7.1.2   Deflation 213
    7.2   Arnoldi Methods 213
    7.3   Linear Model Identification 220
        7.3.1   Prony Method 221
        7.3.2   The Levenberg-Marquardt Method 223
    7.4   Power System Applications 230
        7.4.1   Participation Factors 230
        7.4.2   Modal Analysis 231
    7.5   Problems 235

    References   **237**

    Index   **241**

Chapter 1

Introduction

In today's deregulated environment, the nation's electric power network is being forced to operate in a manner for which it was not intentionally designed. Therefore, system analysis is very important to predict and continually update the operating status of the network. This includes estimating the current power flows and bus voltages (Power Flow Analysis and State Estimation), determining the stability limits of the system (Continuation Power Flow, Numerical Integration for Transient Stability, and Eigenvalue Analysis), and minimizing costs (Optimal Power Flow). This book provides an introductory study of the various computational methods that form the basis of many analytical studies in power systems and other engineering and science fields. This book provides the analytical background of the algorithms used in numerous commercial packages. By understanding the theory behind many of the algorithms, the reader/user can better use the software and make more informed decisions (i.e., choice of integration method and step-size in simulation packages).

Due to the sheer size of the power grid, hand-based calculations are nearly impossible and computers offer the only truly viable means for system analysis. The power industry is one of the largest users of computer technology and one of the first industries to embrace the potential of computer analysis when mainframes first became available. Although the first algorithms for power system analysis were developed in the 1940's, it wasn't until the 1960's when computer usage became widespread within the power industry. Many of the analytical techniques and algorithms used today for the simulation and analysis of large systems were originally developed for power system applications.

As power systems increasingly operate under stressed conditions, computer simulation will play a large role in control and security assessment. Commercial packages routinely fail or give erroneous results when used to simulate stressed systems. Understanding of the underlying numerical algorithms is imperative to correctly interpret the results of commercial packages. For example, will the system really exhibit the simulated behavior or is the simulation simply an artifact of a numerical inaccuracy? The educated user can make better judgments about how to compensate for numerical shortcomings in such packages, either by better choice of simulation parameters or by posing the problem in a more numerically tractable manner. This book will provide the background for a number of widely used numerical algorithms that

1

underlie many commercial packages for power system analysis and design.

This book is intended to be used as a text in conjunction with a semester-long graduate level course in computational algorithms. While the majority of examples in this text are based on power system applications, the theory is presented in a general manner so as to be applicable to a wide range of engineering systems. Although some knowledge of power system engineering may be required to fully appreciate the subtleties of some of the illustrations, such knowledge is not a prerequisite for understanding the algorithms themselves. The text and examples are used to provide an introduction to a wide range of numerical methods without being an exhaustive reference. Many of the algorithms presented in this book have been the subject of numerous modifications and are still the object of on-going research. As this text is intended to provide a foundation, many of these new advances are not explicitly covered, but are rather given as references for the interested reader. The examples in this text are intended to be simple and thorough enough to be reproduced easily. Most "real world" problems are much larger in size and scope, but the methodologies presented in this text should sufficiently prepare the reader to cope with any difficulties he/she may encounter.

Most of the examples in this text were produced using code written in Matlab^{TM}. Although this was the platform used by the author, in practice, any computer language may be used for implementation. There is no practical reason for a preference for any particular platform or language.

Chapter 2

The Solution of Linear Systems

In many branches of engineering and science it is desirable to be able to mathematically determine the state of a system based on a set of physical relationships. These physical relationships may be determined from characteristics such as circuit topology, mass, weight, or force to name a few. For example, the injected currents, network topology, and branch impedances govern the voltages at each node of a circuit. In many cases, the relationship between the known, or input, quantities and the unknown, or output, states is a linear relationship. Therefore, a linear system may be generically modeled as

$$Ax = b \qquad (2.1)$$

where b is the $n \times 1$ vector of known quantities, x is the $n \times 1$ unknown state vector, and A is the $n \times n$ matrix that relates x to b. For the time being, it will be assumed that the matrix A is invertible, or non-singular; thus, each vector b will yield a unique corresponding vector x. Thus the matrix A^{-1} exists and

$$x^* = A^{-1}b \qquad (2.2)$$

is the unique solution to equation (2.1).

The natural approach to solving equation (2.1) is to directly calculate the inverse of A and multiply it by the vector b. One method to calculate A^{-1} is to use *Cramer's* rule :

$$A^{-1}(i,j) = \frac{1}{det(A)}\left(A_{ij}\right)^T \quad \text{for } i = 1,\ldots,n, j = 1,\ldots,n \qquad (2.3)$$

where $A^{-1}(i,j)$ is the ij^{th} entry of A^{-1} and A_{ij} is the cofactor of each entry a_{ij} of A. This method requires the calculation of $(n+1)$ determinants which results in $2(n+1)!$ multiplications to find A^{-1}! For large values of n, the calculation requirement grows too rapidly for computational tractability; thus, alternative approaches have been developed.

Basically there are two approaches to solving equation (2.1):

- The *direct methods*, or elimination methods, find the exact solution (within the accuracy of the computer) through a finite number of arithmetic operations. The solution x of a direct method would be completely accurate were it not for computer roundoff errors.

- *Iterative methods,* on the other hand, generate a sequence of (hopefully) progressively improving approximations to the solution based on the application of the same computational procedure at each step. The iteration is terminated when an approximate solution is obtained having some pre-specified accuracy or when it is determined that the iterates are not improving.

The choice of solution methodology usually relies on the structure of the system under consideration. Certain systems lend themselves more amenably to one type of solution method versus the other. In general, direct methods are best for full matrices, whereas iterative methods are better for matrices that are large and sparse. But as with most generalizations, there are notable exceptions to this rule of thumb.

2.1   Gaussian Elimination

An alternate method for solving equation (2.1) is to solve for x without calculating A^{-1} explicitly. This approach is a *direct method* of linear system solution, since x is found directly. One common direct method is the method of *Gaussian elimination.* The basic idea behind Gaussian elimination is to use the first equation to eliminate the first unknown from the remaining equations. This process is repeated sequentially for the second unknown, the third unknown, etc., until the elimination process is completed. The n-th unknown is then calculated directly from the input vector b. The unknowns are then recursively substituted back into the equations until all unknowns have been calculated.

Gaussian elimination is the process by which the augmented $n \times (n + 1)$ matrix

$$[A \mid b]$$

is converted to the $n \times (n + 1)$ matrix

$$[I \mid b^*]$$

through a series of elementary row operations, where

$$Ax = b$$
$$A^{-1}Ax = A^{-1}b$$
$$Ix = A^{-1}b = b^*$$
$$x^* = b^*$$

Thus if a series of elementary row operations exist that can transform the matrix A into the identity matrix I, then the application of the same set of elementary row operations will also transform the vector b into the solution vector x^*.

An elementary row operation consists of one of three possible actions that can be applied to a matrix:

- interchange any two rows of the matrix

- multiply any row by a constant

- take a linear combination of rows and add it to another row

The elementary row operations are chosen to transform the matrix A into an upper triangular matrix that has ones on the diagonal and zeros in the sub-diagonal positions. This process is known as the *forward elimination* step. Each step in the forward elimination can be obtained by successively pre-multiplying the matrix A by an elementary matrix ξ, where ξ is the matrix obtained by performing an elementary row operation on the identity matrix.

Example 2.1
Find a sequence of elementary matrices that when applied to the following matrix will produce an upper triangular matrix.

$$A = \begin{bmatrix} 1 & 3 & 4 & 8 \\ 2 & 1 & 2 & 3 \\ 4 & 3 & 5 & 8 \\ 9 & 2 & 7 & 4 \end{bmatrix}$$

Solution 2.1 To upper triangularize the matrix, the elementary row operations will need to systematically zero out each column below the diagonal. This can be achieved by replacing each row of the matrix below the diagonal with the difference of the row itself and a constant times the diagonal row, where the constant is chosen to result in a zero sum in the column under the diagonal. Therefore row 2 of A is replaced by (row 2 - 2(row 1)) and the elementary matrix is

$$\xi_1 = \begin{bmatrix} 1 & 0 & 0 & 0 \\ -2 & 1 & 0 & 0 \\ 0 & 0 & 1 & 0 \\ 0 & 0 & 0 & 1 \end{bmatrix}$$

and

$$\xi_1 A = \begin{bmatrix} 1 & 3 & 4 & 8 \\ 0 & -5 & -6 & -13 \\ 4 & 3 & 5 & 8 \\ 9 & 2 & 7 & 4 \end{bmatrix}$$

Note that all rows except row 2 remain the same and row 2 now has a 0 in the column under the first diagonal. Similarly the two elementary matrices that complete the elimination of the first column are:

$$\xi_2 = \begin{bmatrix} 1 & 0 & 0 & 0 \\ 0 & 1 & 0 & 0 \\ -4 & 0 & 1 & 0 \\ 0 & 0 & 0 & 1 \end{bmatrix}$$

$$\xi_3 = \begin{bmatrix} 1 & 0 & 0 & 0 \\ 0 & 1 & 0 & 0 \\ 0 & 0 & 1 & 0 \\ -9 & 0 & 0 & 1 \end{bmatrix}$$

and

$$\xi_3 \xi_2 \xi_1 A = \begin{bmatrix} 1 & 3 & 4 & 8 \\ 0 & -5 & -6 & -13 \\ 0 & -9 & -11 & -24 \\ 0 & -25 & -29 & -68 \end{bmatrix} \tag{2.4}$$

The process is now applied to the second column to zero out everything below the second diagonal and scale the diagonal to one. Therefore

$$\xi_4 = \begin{bmatrix} 1 & 0 & 0 & 0 \\ 0 & 1 & 0 & 0 \\ 0 & -\frac{9}{5} & 1 & 0 \\ 0 & 0 & 0 & 1 \end{bmatrix}$$

$$\xi_5 = \begin{bmatrix} 1 & 0 & 0 & 0 \\ 0 & 1 & 0 & 0 \\ 0 & 0 & 1 & 0 \\ 0 & -\frac{25}{5} & 0 & 1 \end{bmatrix}$$

$$\xi_6 = \begin{bmatrix} 1 & 0 & 0 & 0 \\ 0 & -\frac{1}{5} & 0 & 0 \\ 0 & 0 & 1 & 0 \\ 0 & 0 & 0 & 1 \end{bmatrix}$$

Similarly,

$$\xi_6 \xi_5 \xi_4 \xi_3 \xi_2 \xi_1 A = \begin{bmatrix} 1 & 3 & 4 & 8 \\ 0 & 1 & \frac{6}{5} & \frac{13}{5} \\ 0 & 0 & -\frac{1}{5} & -\frac{3}{5} \\ 0 & 0 & 1 & -3 \end{bmatrix} \tag{2.5}$$

Similarly,

$$\xi_7 = \begin{bmatrix} 1 & 0 & 0 & 0 \\ 0 & 1 & 0 & 0 \\ 0 & 0 & 1 & 0 \\ 0 & 0 & 5 & 1 \end{bmatrix}$$

$$\xi_8 = \begin{bmatrix} 1 & 0 & 0 & 0 \\ 0 & 1 & 0 & 0 \\ 0 & 0 & -5 & 0 \\ 0 & 0 & 0 & 1 \end{bmatrix}$$

yielding

$$\xi_8\xi_7\xi_6\xi_5\xi_4\xi_3\xi_2\xi_1 A = \begin{bmatrix} 1 & 3 & 4 & 8 \\ 0 & 1 & \frac{6}{5} & \frac{13}{5} \\ 0 & 0 & 1 & 3 \\ 0 & 0 & 0 & -6 \end{bmatrix} \tag{2.6}$$

Lastly,

$$\xi_9 = \begin{bmatrix} 1 & 0 & 0 & 0 \\ 0 & 1 & 0 & 0 \\ 0 & 0 & 1 & 0 \\ 0 & 0 & 0 & -\frac{1}{6} \end{bmatrix}$$

and

$$\xi_9\xi_8\xi_7\xi_6\xi_5\xi_4\xi_3\xi_2\xi_1 A = \begin{bmatrix} 1 & 3 & 4 & 8 \\ 0 & 1 & \frac{6}{5} & \frac{13}{5} \\ 0 & 0 & 1 & 3 \\ 0 & 0 & 0 & 1 \end{bmatrix} \tag{2.7}$$

which completes the upper triangularization process. ∎

Once an upper triangular matrix has been achieved, the solution vector x^* can be found by successive substitution (or *back substitution*) of the states.

Example 2.2

Using the upper triangular matrix of Example 2.1, find the solution to

$$\begin{bmatrix} 1 & 3 & 4 & 8 \\ 2 & 1 & 2 & 3 \\ 4 & 3 & 5 & 8 \\ 9 & 2 & 7 & 4 \end{bmatrix} \begin{bmatrix} x_1 \\ x_2 \\ x_3 \\ x_4 \end{bmatrix} = \begin{bmatrix} 1 \\ 1 \\ 1 \\ 1 \end{bmatrix}$$

Solution 2.2 Note that the product of a series of lower triangular matrices is lower triangular; therefore, the product

$$W = \xi_9\xi_8\xi_7\xi_6\xi_5\xi_4\xi_3\xi_2\xi_1 \tag{2.8}$$

is lower triangular. Since the application of the elementary matrices to the matrix A results in an upper triangular matrix, then

$$WA = U \tag{2.9}$$

where U is the upper triangular matrix that results from the forward elimination process. Premultiplying equation (2.1) by W yields

$$W A x = W b \tag{2.10}$$
$$U x = W b \tag{2.11}$$
$$= b' \tag{2.12}$$

where $W b = b'$.

From Example 2.1:

$$W = \begin{bmatrix} 1 & 0 & 0 & 0 \\ \frac{2}{5} & -\frac{1}{5} & 0 & 0 \\ 2 & 9 & -5 & 0 \\ \frac{1}{6} & \frac{14}{6} & -\frac{5}{6} & -\frac{1}{6} \end{bmatrix}$$

and

$$b' = W \begin{bmatrix} 1 \\ 1 \\ 1 \\ 1 \end{bmatrix} = \begin{bmatrix} 1 \\ \frac{1}{5} \\ 6 \\ \frac{3}{2} \end{bmatrix}$$

Thus,

$$\begin{bmatrix} 1 & 3 & 4 & 8 \\ 0 & 1 & \frac{6}{5} & \frac{13}{5} \\ 0 & 0 & 1 & 3 \\ 0 & 0 & 0 & 1 \end{bmatrix} \begin{bmatrix} x_1 \\ x_2 \\ x_3 \\ x_4 \end{bmatrix} = \begin{bmatrix} 1 \\ \frac{1}{5} \\ 6 \\ \frac{3}{2} \end{bmatrix} \tag{2.13}$$

By inspection, $x_4 = \frac{3}{2}$. The third row yields

$$x_3 = 6 - 3x_4 \tag{2.14}$$

Substituting the value of x_4 into equation (2.14) yields $x_3 = \frac{3}{2}$. Similarly,

$$x_2 = \frac{1}{5} - \frac{6}{5}x_3 - \frac{13}{5}x_4 \tag{2.15}$$

and substituting x_3 and x_4 into equation (2.15) yields $x_2 = -\frac{11}{2}$. Solving for x_1 in a similar manner produces

$$x_1 = 1 - 3x_2 - 4x_3 - 8x_4 \tag{2.16}$$
$$= -\frac{1}{2} \tag{2.17}$$

Thus,

$$\begin{bmatrix} x_1 \\ x_2 \\ x_3 \\ x_4 \end{bmatrix} = \frac{1}{2} \begin{bmatrix} -1 \\ -11 \\ 3 \\ 3 \end{bmatrix} \quad \blacksquare$$

The solution methodology of successively substituting values of x back into the equation as they are found gives rise to the name of *back substitution* for this step of the Gaussian elimination. Therefore, Gaussian elimination consists of two main steps: forward elimination and back substitution. Forward elimination is the process of transforming the matrix A into triangular factors. Back substitution is the process by which the unknown vector x is found from the input vector b and the factors of A. Gaussian elimination also provides the framework under which the LU factorization process is developed.

2.2   LU Factorization

The forward elimination step of Gaussian elimination produces a series of upper and lower triangular matrices that are related to the A matrix as given in equation (2.9). The matrix W is a lower triangular matrix and U is an upper triangular matrix with ones on the diagonal. Recall that the inverse of a lower triangular matrix is also a lower triangular matrix; therefore, if

$$L \triangleq W^{-1}$$

then

$$A = LU$$

The matrices L and U give rise to the name of the factorization/elimination algorithm known as "LU factorization." In fact, given any nonsingular matrix A, there exists some permutation matrix P (possibly $P = I$), such that

$$LU = PA \tag{2.18}$$

where U is upper triangular with unit diagonals, L is lower triangular with nonzero diagonals, and P is a matrix of ones and zeros obtained by rearranging the rows and columns of the identity matrix. Once a proper matrix P is chosen, this factorization is unique [3]. Once $P, L,$ and U are determined, then the system

$$Ax = b \tag{2.19}$$

can be solved expeditiously. Premultiplying equation (2.19) by the matrix P yields

$$PAx = Pb = b' \tag{2.20}$$

$$LUx = b' \tag{2.21}$$

where b' is just a rearrangement of the vector b. Introducing a "dummy" vector y such that

$$Ux = y \tag{2.22}$$

thus

$$Ly = b' \qquad (2.23)$$

Consider the structure of equation (2.23):

$$\begin{bmatrix} l_{11} & 0 & 0 & \cdots & 0 \\ l_{21} & l_{22} & 0 & \cdots & 0 \\ l_{31} & l_{32} & l_{33} & \cdots & 0 \\ \vdots & \vdots & \vdots & \ddots & \vdots \\ l_{n1} & l_{n2} & l_{n3} & \cdots & l_{nn} \end{bmatrix} \begin{bmatrix} y_1 \\ y_2 \\ y_3 \\ \vdots \\ y_n \end{bmatrix} = \begin{bmatrix} b'_1 \\ b'_2 \\ b'_3 \\ \vdots \\ b'_n \end{bmatrix}$$

The elements of the vector y can be found by straightforward substitution:

$$y_1 = \frac{b'_1}{l_{11}}$$

$$y_2 = \frac{1}{l_{22}} (b'_2 - l_{21} y_1)$$

$$y_3 = \frac{1}{l_{33}} (b'_3 - l_{31} y_1 - l_{32} y_2)$$

$$\vdots$$

$$y_n = \frac{1}{l_{nn}} \left(b'_n - \sum_{j=1}^{n-1} l_{nj} y_j \right)$$

After the vector y has been found, then x can be easily found from

$$\begin{bmatrix} 1 & u_{12} & u_{13} & \cdots & u_{1n} \\ 0 & 1 & u_{23} & \cdots & u_{2n} \\ 0 & 0 & 1 & \cdots & u_{3n} \\ \vdots & \vdots & \vdots & \ddots & \vdots \\ 0 & 0 & 0 & \vdots & 1 \end{bmatrix} \begin{bmatrix} x_1 \\ x_2 \\ x_3 \\ \vdots \\ x_n \end{bmatrix} = \begin{bmatrix} y_1 \\ y_2 \\ y_3 \\ \vdots \\ y_n \end{bmatrix}$$

Similarly, the solution vector x can be found by backward substitution:

$$x_n = y_n$$

$$x_{n-1} = y_{n-1} - u_{n-1,n} x_n$$

$$x_{n-2} = y_{n-2} - u_{n-2,n} x_n - u_{n-2,n-1} x_{n-1}$$

$$\vdots$$

$$x_1 = y_1 - \sum_{j=2}^{n} u_{1j} x_j$$

The value of LU factorization is that once A is factored into the upper and lower triangular matrices, the solution for the solution vector x is straightforward. Note that the inverse to A is never explicitly found.

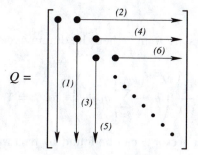

$$Q =$$

FIGURE 2.1
Order of calculating columns and rows of Q

Several methods for computing the LU factors exist and each method has its advantages and disadvantages. One common factorization approach is known as the *Crout's* algorithm for finding the LU factors [3]. Let the matrix Q be defined as

$$Q \triangleq L + U - I = \begin{bmatrix} l_{11} & u_{12} & u_{13} & \cdots & u_{1n} \\ l_{21} & l_{22} & u_{23} & \cdots & u_{2n} \\ l_{31} & l_{32} & l_{33} & \cdots & u_{3n} \\ \vdots & \vdots & \vdots & \ddots & \vdots \\ l_{n1} & l_{n2} & l_{n3} & \cdots & l_{nn} \end{bmatrix} \quad (2.24)$$

Crout's algorithm computes the elements of Q first by column and then row as shown in Figure 2.1. Each element q_{ij} of Q depends only on the a_{ij} entry of A and previously computed values of Q.

Crout's Algorithm for Computing LU from A

1. Initialize Q to the zero matrix. Let $j = 1$.

2. Complete the jth column of Q (j^{th} column of L) as

$$q_{kj} = a_{kj} - \sum_{i=1}^{j-1} q_{ki}q_{ij} \quad \text{for } k = j, \ldots, n \quad (2.25)$$

3. If $j = n$, then stop.

4. Assuming that $q_{jj} \neq 0$, complete the jth row of Q (j^{th} row of U) as

$$q_{jk} = \frac{1}{q_{jj}} \left(a_{jk} - \sum_{i=1}^{j-1} q_{ji}q_{ik} \right) \quad \text{for } k = j+1, \ldots, n \quad (2.26)$$

5. Set $j = j + 1$. Go to step 2.

Once the *LU* factors are found, then the dummy vector y can be found by forward substitution:

$$y_k = \frac{1}{q_{kk}} \left(b_k - \sum_{j=1}^{k-1} q_{kj} y_j \right) \quad \text{for } k = 1, \ldots, n \qquad (2.27)$$

Similarly, the solution vector x can be found by backward substitution:

$$x_k = y_k - \sum_{j=k+1}^{n} q_{kj} x_j \quad \text{for } k = n, n-1, \ldots, 1 \qquad (2.28)$$

One measure of the computation involved in the LU factorization process is to count the number of multiplications and divisions required to find the solution since these are both floating point operations. Computing the jth column of Q (j^{th} column of L) requires

$$\sum_{j=1}^{n} \sum_{k=j}^{n} (j-1)$$

multiplications and divisions. Similarly, computing the jth row of Q (j^{th} row of U) requires

$$\sum_{j=1}^{n-1} \sum_{k=j+1}^{n} j$$

multiplications and divisions. The forward substitution step requires

$$\sum_{j=1}^{n} j$$

and the backward substitution step requires

$$\sum_{j=1}^{n} (n-j)$$

multiplications and divisions. Taken together, the LU factorization procedure requires

$$\frac{1}{3} \left(n^3 - n \right)$$

and the substitution steps require n^2 multiplications and divisions. Therefore the whole process of solving the linear system of equation (2.1) requires a total of

$$\frac{1}{3} \left(n^3 - n \right) + n^2 \qquad (2.29)$$

multiplications and divisions. Compare this to the requirements of Cramer's rule which requires $2(n+1)!$ multiplications and divisions. Obviously for a system of any significant size, it is far more computationally efficient to use LU factorization and forward/backward substitution to find the solution x.

Example 2.3

Using LU factorization with forward and backward substitution, find the solution to the system of Example 2.2.

Solution 2.3 The first step is to find the LU factors of the A matrix:

$$A = \begin{bmatrix} 1 & 3 & 4 & 8 \\ 2 & 1 & 2 & 3 \\ 4 & 3 & 5 & 8 \\ 9 & 2 & 7 & 4 \end{bmatrix}$$

Starting with $j = 1$, equation (2.25) indicates that the elements of the first column of Q are identical to the elements of the first column of A. Similarly, according to equation (2.26), the first row of Q becomes:

$$q_{12} = \frac{a_{12}}{q_{11}} = \frac{3}{1} = 3$$

$$q_{13} = \frac{a_{13}}{q_{11}} = \frac{4}{1} = 4$$

$$q_{14} = \frac{a_{14}}{q_{11}} = \frac{8}{1} = 8$$

Thus for $j = 1$, the Q matrix becomes:

$$Q = \begin{bmatrix} 1 & 3 & 4 & 8 \\ 2 & & & \\ 4 & & & \\ 9 & & & \end{bmatrix}$$

For $j = 2$, the second column and row of Q below and to the right of the diagonal, respectively, will be calculated. For the second column of Q:

$$q_{22} = a_{22} - q_{21}q_{12} = 1 - (2)(3) = -5$$
$$q_{32} = a_{32} - q_{31}q_{12} = 3 - (4)(3) = -9$$
$$q_{42} = a_{42} - q_{41}q_{12} = 2 - (9)(3) = -25$$

Each element of Q uses the corresponding element of A and elements of Q that have been previously computed. Note also that the inner indices of the products are always the same and the outer indices are the same as the indices of the element being computed. This holds true for both column and row calculations. The second row of Q is computed:

$$q_{23} = \frac{1}{q_{22}}(a_{23} - q_{21}q_{13}) = \frac{1}{-5}(2 - (2)(4)) = \frac{6}{5}$$

$$q_{24} = \frac{1}{q_{22}}(a_{24} - q_{21}q_{14}) = \frac{1}{-5}(3 - (2)(8)) = \frac{13}{5}$$

After $j = 2$, the Q matrix becomes:

$$Q = \begin{bmatrix} 1 & 3 & 4 & 8 \\ 2 & -5 & \frac{6}{5} & \frac{13}{5} \\ 4 & -9 & & \\ 9 & -25 & & \end{bmatrix}$$

Continuing on for $j = 3$, the third column of Q is calculated

$$q_{33} = a_{33} - (q_{31}q_{13} + q_{32}q_{23}) = 5 - \left((4)(4) + (-9)\frac{6}{5}\right) = -\frac{1}{5}$$

$$q_{43} = a_{43} - (q_{41}q_{13} + q_{42}q_{23}) = 7 - \left((9)(4) + (-25)\frac{6}{5}\right) = 1$$

and the third row of Q becomes

$$q_{34} = \frac{1}{q_{33}}(a_{34} - (q_{31}q_{14} + q_{32}q_{24}))$$

$$= (-5)\left(8 - \left((4)(8) + (-9)\left(\frac{13}{5}\right)\right)\right) = 3$$

yielding

$$Q = \begin{bmatrix} 1 & 3 & 4 & 8 \\ 2 & -5 & \frac{6}{5} & \frac{13}{5} \\ 4 & -9 & -\frac{1}{5} & 1 \\ 9 & -25 & 3 & \end{bmatrix}$$

Lastly, for $j = 4$, the final diagonal element is found:

$$q_{44} = a_{44} - (q_{41}q_{14} + q_{42}q_{24} + q_{43}q_{34})$$

$$= 4 - \left((9)(8) + (-25)\left(\frac{13}{5}\right) + (3)(1)\right) = -6$$

Thus:

$$Q = \begin{bmatrix} 1 & 3 & 4 & 8 \\ 2 & -5 & \frac{6}{5} & \frac{13}{5} \\ 4 & -9 & -\frac{1}{5} & 3 \\ 9 & -25 & 1 & -6 \end{bmatrix}$$

$$L = \begin{bmatrix} 1 & 0 & 0 & 0 \\ 2 & -5 & 0 & 0 \\ 4 & -9 & -\frac{1}{5} & 0 \\ 9 & -25 & 1 & -6 \end{bmatrix}$$

$$U = \begin{bmatrix} 1 & 3 & 4 & 8 \\ 0 & 1 & \frac{6}{5} & \frac{13}{5} \\ 0 & 0 & 1 & 3 \\ 0 & 0 & 0 & 1 \end{bmatrix}$$

One method of checking the correctness of the solution is to check if $LU = A$, which in this case it does.

Once the LU factors have been found, then the next step in the solution process is the forward elimination using the L matrix and the b vector to find the dummy vector y. Using forward substitution to solve $Ly = b$ for y:

$$y_1 = \frac{b_1}{L_{11}} = \frac{1}{1} = 1$$

$$y_2 = \frac{(b_2 - L_{21}y_1)}{L_{22}} = \frac{(1 - (2)(1))}{-5} = \frac{1}{5}$$

$$y_3 = \frac{(b_3 - (L_{31}y_1 + L_{32}y_2))}{L_{33}} = (-5)\left(1 - \left((4)(1) + (-9)\frac{1}{5}\right)\right) = 6$$

$$y_4 = \frac{(b_4 - (L_{41}y_1 + L_{42}y_2 + L_{43}y_3))}{L_{44}}$$

$$= \frac{\left(1 - ((9)(1) + (-25)\left(\frac{1}{5}\right) + (1)(6))\right)}{-6} = \frac{3}{2}$$

Thus

$$y = \begin{bmatrix} 1 \\ \frac{1}{5} \\ 6 \\ \frac{3}{2} \end{bmatrix}$$

Similarly, backward substitution is then applied to $Ux = y$ to find the solution vector x:

$$x_4 = y_4 = \frac{3}{2}$$

$$x_3 = y_3 - U_{34}x_4 = 6 - (3)\left(\frac{3}{2}\right) = \frac{3}{2}$$

$$x_2 = y_2 - (U_{24}x_4 + U_{23}x_3) = \frac{1}{5} - \left(\left(\frac{13}{5}\right)\left(\frac{3}{2}\right) + \left(\frac{6}{5}\right)\left(\frac{3}{2}\right)\right) = -\frac{11}{2}$$

$$x_1 = y_1 - (U_{14}x_4 + U_{13}x_3 + U_{12}x_2)$$

$$= 1 - \left((8)\left(\frac{3}{2}\right) + (4)\left(\frac{3}{2}\right) + (3)\left(-\frac{11}{2}\right)\right) = -\frac{1}{2}$$

yielding the final solution vector

$$x = \frac{1}{2} \begin{bmatrix} -1 \\ -11 \\ 3 \\ 3 \end{bmatrix}$$

which is the same solution found by Gaussian elimination and backward substitution in Example 2.2. A quick check to verify the correctness of the solution is to substitute the solution vector x back into the linear system $Ax = b$. ∎

2.2.1   LU Factorization with Partial Pivoting

The LU factorization process presented assumes that the diagonal element is non-zero. Not only must the diagonal element be non-zero, it must be on the same order of magnitude as the other non-zero elements. Consider the solution of the following linear system:

$$\begin{bmatrix} 10^{-10} & 1 \\ 2 & 1 \end{bmatrix} \begin{bmatrix} x_1 \\ x_2 \end{bmatrix} = \begin{bmatrix} 1 \\ 5 \end{bmatrix} \tag{2.30}$$

By inspection, the solution to this linear system is

$$x_1 \approx 2$$
$$x_2 \approx 1$$

The LU factors for A are

$$L = \begin{bmatrix} 10^{-10} & 0 \\ 2 & (1 - 2 \times 10^{10}) \end{bmatrix}$$
$$U = \begin{bmatrix} 1 & 10^{10} \\ 0 & 1 \end{bmatrix}$$

Applying forward elimination to solve for the dummy vector y yields:

$$y_1 = 10^{10}$$
$$y_2 = \frac{(5 - 2 \times 10^{10})}{(1 - 2 \times 10^{10})} \approx 1$$

Back substituting y into $Ux = y$ yields

$$x_2 = y_2 \approx 1$$
$$x_1 = 10^{10} - 10^{10} x_2 \approx 0$$

The solution for x_2 is correct, but the solution for x_1 is way off. Why did this happen? The problem with the equations arranged the way they are in equation (2.30) is that 10^{-10} is too near zero for most computers. However, if the equations are rearranged such that

$$\begin{bmatrix} 2 & 1 \\ 10^{-10} & 1 \end{bmatrix} \begin{bmatrix} x_1 \\ x_2 \end{bmatrix} = \begin{bmatrix} 5 \\ 1 \end{bmatrix} \qquad (2.31)$$

then the LU factors become

$$L = \begin{bmatrix} 2 & 0 \\ 10^{-10} & 1 - \frac{1}{2} \times 10^{-10} \end{bmatrix}$$

$$U = \begin{bmatrix} 1 & \frac{1}{2} \\ 0 & 1 \end{bmatrix}$$

The dummy vector y becomes

$$y_1 = \frac{5}{2}$$

$$y_2 = \frac{\left(1 - \frac{5}{2} \times 10^{-10}\right)}{\left(1 - \frac{1}{2} \times 10^{-10}\right)} \approx 1$$

and by back substitution, x becomes

$$x_2 \approx 1$$

$$x_1 \approx \frac{5}{2} - \frac{1}{2}(1) = 2$$

which is the solution obtained by inspection of the equations. Therefore even though the diagonal entry may not be exactly zero, it is still good practice to rearrange the equations such that the largest magnitude element lies on the diagonal. This process is known as *pivoting* and gives rise to the permutation matrix P of equation (2.18).

Since the Crout's algorithm computes the Q matrix by column and row with increasing index, only *partial pivoting* can used, that is, only the rows of Q (and correspondingly A) can be exchanged. The columns must remain static. To choose the best pivot, the column beneath the jth diagonal (at the jth step in the LU factorization) is searched for the element with the largest absolute value. The corresponding row and the jth row are then exchanged. The pivoting strategy may be succinctly expressed as:

Partial Pivoting Strategy

1. At the jth step of LU factorization, choose the kth row as the exchange row such that

$$|q_{jj}| = max\,|q_{kj}| \text{ for } k = j, \ldots, n \qquad (2.32)$$

2. Exchange rows and update A, P, and Q correspondingly.

The permutation matrix P is comprised of ones and zeros and is obtained as the product of a series of elementary permutation matrices $P^{j,k}$ which represent the exchange of rows j and k. The elementary permutation matrix $P^{j,k}$, shown in Figure 2.2, is obtained from the identify matrix by interchanging rows j and k. A pivot is achieved by the pre-multiplication of a properly chosen $P^{j,k}$. Since this is only an interchange of rows, the order of the unknown vector does not change.

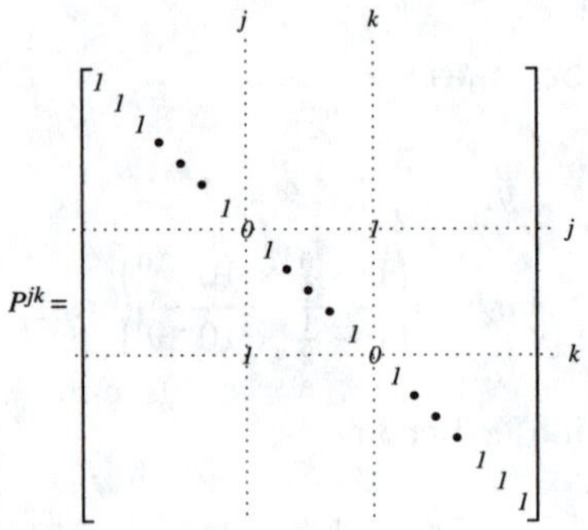

FIGURE 2.2
Elementary permutation matrix $P^{j,k}$

Example 2.4

Repeat Example 2.3 using partial pivoting.

Solution 2.4 The A matrix is repeated here for convenience.

$$A = \begin{bmatrix} 1 & 3 & 4 & 8 \\ 2 & 1 & 2 & 3 \\ 4 & 3 & 5 & 8 \\ 9 & 2 & 7 & 4 \end{bmatrix}$$

For $j = 1$, the first column of Q is exactly the first column of A. Applying the pivoting strategy of equation (2.32), the q_{41} element has the largest magnitude of the first column; therefore, rows four and one are exchanged. The elementary permutation matrix $P^{1,4}$ is

$$P^{1,4} = \begin{bmatrix} 0 & 0 & 0 & 1 \\ 0 & 1 & 0 & 0 \\ 0 & 0 & 1 & 0 \\ 1 & 0 & 0 & 0 \end{bmatrix}$$

The corresponding A matrix becomes

$$A = \begin{bmatrix} 9 & 2 & 7 & 4 \\ 2 & 1 & 2 & 3 \\ 4 & 3 & 5 & 8 \\ 1 & 3 & 4 & 8 \end{bmatrix}$$

and Q at the $j = 1$ step:

$$Q = \begin{bmatrix} 9 & \frac{2}{9} & \frac{7}{9} & \frac{4}{9} \\ 2 & & & \\ 4 & & & \\ 1 & & & \end{bmatrix}$$

At $j = 2$, the calculation of the second column of Q yields

$$Q = \begin{bmatrix} 9 & \frac{2}{9} & \frac{7}{9} & \frac{4}{9} \\ 2 & \frac{5}{9} & & \\ 4 & \frac{19}{9} & & \\ 1 & \frac{25}{9} & & \end{bmatrix}$$

Searching the elements in the jth column below the diagonal, the fourth row of the jth (i.e., second) column once again yields the largest magnitude. Therefore rows two and four must be exchanged, yielding the elementary permutation matrix $P^{2,4}$:

$$P^{2,4} = \begin{bmatrix} 1 & 0 & 0 & 0 \\ 0 & 0 & 0 & 1 \\ 0 & 0 & 1 & 0 \\ 0 & 1 & 0 & 0 \end{bmatrix}$$

Similarly, the updated A is

$$\begin{bmatrix} 9 & 2 & 7 & 4 \\ 1 & 3 & 4 & 8 \\ 4 & 3 & 5 & 8 \\ 2 & 1 & 2 & 3 \end{bmatrix}$$

which yields the following Q:

$$Q = \begin{bmatrix} 9 & \frac{2}{9} & \frac{7}{9} & \frac{4}{9} \\ 1 & \frac{25}{9} & \frac{29}{25} & \frac{68}{25} \\ 4 & \frac{19}{9} & & \\ 2 & \frac{5}{9} & & \end{bmatrix}$$

For $j = 3$, the calculation of the third column of Q yields:

$$Q = \begin{bmatrix} 9 & \frac{2}{9} & \frac{7}{9} & \frac{4}{9} \\ 1 & \frac{25}{9} & \frac{29}{25} & \frac{68}{25} \\ 4 & \frac{19}{9} & -\frac{14}{25} & \\ 2 & \frac{5}{9} & -\frac{1}{5} & \end{bmatrix}$$

In this case, the diagonal element has the largest magnitude, so no pivoting is required. Continuing with the calculation of the 3rd row of Q yields:

$$Q = \begin{bmatrix} 9 & \frac{2}{9} & \frac{7}{9} & \frac{4}{9} \\ 1 & \frac{25}{9} & \frac{29}{25} & \frac{68}{25} \\ 4 & \frac{19}{9} & -\frac{14}{25} & -\frac{13}{14} \\ 2 & \frac{5}{9} & -\frac{1}{5} & \end{bmatrix}$$

Lastly, calculating q_{44} yields the final Q matrix:

$$Q = \begin{bmatrix} 9 & \frac{2}{9} & \frac{7}{9} & \frac{4}{9} \\ 1 & \frac{25}{9} & \frac{29}{25} & \frac{68}{25} \\ 4 & \frac{19}{9} & -\frac{14}{25} & -\frac{13}{14} \\ 2 & \frac{5}{9} & -\frac{1}{5} & \frac{3}{7} \end{bmatrix}$$

The permutation matrix P is found by multiplying together the two elementary permutation matrices:

$$P = P^{2,4} P^{1,4} I$$
$$= \begin{bmatrix} 0 & 0 & 0 & 1 \\ 1 & 0 & 0 & 0 \\ 0 & 0 & 1 & 0 \\ 0 & 1 & 0 & 0 \end{bmatrix}$$

The results can be checked to verify that $PA = LU$. The forward and backward substitution steps are carried out on the modified vector $b' = Pb$. ∎

2.2.2   LU Factorization with Complete Pivoting

An alternate LU factorization that allows complete pivoting is the *Gauss'* method. In this approach, two permutation matrices are developed: one for row exchange as in partial pivoting, and a second matrix for column exchange. In this approach, the LU factors are found such that

$$P_1 A P_2 = LU \tag{2.33}$$

Therefore to solve the linear system of equations $Ax = b$ requires that a slightly different approach be used. As with partial pivoting, the permutation matrix P_1 premultiplies the linear system:

$$P_1 Ax = P_1 b = b' \tag{2.34}$$

Now, define a new vector z such that

$$x = P_2 z \tag{2.35}$$

Then substituting equation (2.35) into equation (2.34) yields

$$P_1 A P_2 z = P_1 b = b'$$
$$L U z = b' \tag{2.36}$$

where equation (2.36) can be solved using forward and backward substitution for z. Once z is obtained, then the solution vector x follows from equation (2.35).

In complete pivoting, both rows and columns may be interchanged to place the largest element (in magnitude) on the diagonal at each step in the LU factorization process. The pivot element is chosen from the remaining elements below and to the right of the diagonal.

Complete Pivoting Strategy

1. At the jth step of LU factorization, choose the pivot element such that

$$|q_{jj}| = max\,|q_{kl}|\ \text{for}\ k = j, \ldots, n, \text{and}\ l = j, \ldots, n \tag{2.37}$$

2. Exchange rows and update A, P, and Q correspondingly.

Gauss' Algorithm for Computing LU from A

1. Initialize Q to the zero matrix. Let $j = 1$.

2. Set the jth column of Q (j^{th} column of L) to the jth column of the reduced matrix $A^{(j)}$, where $A^{(1)} = A$, and

$$q_{kj} = a_{kj}^{(j)}\ \text{for}\ k = j, \ldots, n \tag{2.38}$$

3. If $j = n$, then stop.

4. Assuming that $q_{jj} \neq 0$, set the jth row of Q (j^{th} row of U) as

$$q_{jk} = \frac{a_{jk}^{(j)}}{q_{jj}}\ \text{for}\ k = j+1, \ldots, n \tag{2.39}$$

5. Update $A^{(j+1)}$ from $A^{(j)}$ as

$$a_{ik}^{(j+1)} = a_{ik}^{(j)} - q_{ij} q_{jk}\ \text{for}\ i = j+1, \ldots, n, \text{and}\ k = j+1, \ldots, n \tag{2.40}$$

6. Set $j = j + 1$. Go to step 2.

This factorization algorithm gives rise to the same number of multiplications and divisions as Crout's algorithm for LU factorization. Crout's algorithm uses each entry of the A matrix only once, whereas Gauss' algorithm updates the A matrix each time. One advantage of Crout's algorithm over Gauss' algorithm is each element of the A matrix is used only once. Since each q_{jk} is a function of a_{jk} and then a_{jk} is never used again, the element q_{jk} can be written *over* the a_{jk} element. Therefore, rather than having to store two $n \times n$ matrices in memory (A and Q), only one matrix is required.

The Crout's and Gauss' algorithms are only two of numerous algorithms for LU factorization. Other methods include Doolittle and bifactorization algorithms [14], [18], [36]. Most of these algorithms require similar numbers of multiplications and divisions and only differ slightly in performance when implemented on traditional serial computers. However, these algorithms differ considerably when factors such as memory access, storage, and parallelization are considered. Consequently, it is wise to choose the factorization algorithm to fit the application and the computer architecture upon which it will be implemented.

2.3   Condition Numbers and Error Propagation

The Gaussian elimination and LU factorization algorithms are considered direct methods because they calculate the solution vector $x^* = A^{-1}b$ in a finite number of steps without an iterative refinement. On a computer with infinite precision, direct methods would yield the exact solution x^*. However, since computers have finite precision, the solution obtained has limited accuracy. The *condition number* of a matrix is a useful measure for determining the level of accuracy of a solution. The condition number of the matrix A is generally defined as

$$\kappa\left(A\right) = \sqrt{\frac{\lambda_{max}}{\lambda_{min}}} \tag{2.41}$$

where λ_{max} and λ_{min} denote the largest and smallest eigenvalues of the matrix $A^T A$. These eigenvalues are real and non-negative regardless of whether the eigenvalues of A are real or complex.

The condition number of a matrix is a measure of the linear independence of the eigenvectors of the matrix. A singular matrix has at least one zero eigenvalue and contains at least one degenerate row (i.e., the row can be expressed as a linear combination of other rows). The identity matrix, which gives rise to the most linearly independent eigenvectors possible and has every eigenvalue equal to one, has a condition number of 1. If the condition number of a matrix is much much greater than one, then the matrix is said to be *ill conditioned*. The larger the condition number, the more sensitive the solution process is to slight perturbations in the elements of A and the more numerical error likely to be contained in the solution.

Because of numerical error introduced into the solution process, the computed solution $\tilde{x}$ of equation (2.1) will differ from the exact solution x^* by a finite amount Δx. Other errors, such as approximation, measurement, or round-off error, may be introduced into the matrix A and vector b. Gaussian elimination produces a solution that has roughly

$$t \log_{10} \beta - \log_{10} \kappa(A) \tag{2.42}$$

correct decimal places in the solution, where t is the bit length of the mantissa ($t = 24$ for a typical 32-bit binary word), β is the base ($\beta = 2$ for binary operations), and κ is the condition number of the matrix A. One interpretation of equation (2.42) is that the solution will lose about $\log_{10} \kappa$ digits of accuracy during Gaussian elimination (and consequently LU factorization). Based upon the known accuracy of the matrix entries, the condition number, and the machine precision, the accuracy of the numerical solution $\tilde{x}$ can be predicted [27].

2.4 Relaxation Methods

Relaxation methods are iterative in nature and produce a sequence of vectors that ideally converge to the solution $x^* = A^{-1}b$. Relaxation methods can be incorporated into the solution of equation (2.1) in several ways. In all cases, the principal advantage of using a relaxation method stems from not requiring a direct solution of a large system of linear equations and from the fact that the relaxation methods permit the simulator to exploit the latent portions of the system (those portions which are relatively unchanging at the present time) effectively. In addition, with the advent of parallel-processing technology, relaxation methods lend themselves more readily to parallel implementation than do direct methods. The two most common relaxation methods are the Jacobi and the Gauss-Seidel methods [43].

These relaxation methods may be applied for the solution of the linear system

$$Ax = b \tag{2.43}$$

A general approach to relaxation methods is to define a *splitting matrix M* such that equation (2.43) can be rewritten in equivalent form as

$$Mx = (M - A)x + b \tag{2.44}$$

This splitting leads to the iterative process

$$Mx^{k+1} = (M - A)x^k + b \quad k = 1, \dots, \infty \tag{2.45}$$

where k is the iteration index. This iteration produces a sequence of vectors $x^1, x^2, \ldots$ for a given initial guess x^0. Various iterative methods can be developed by different choices of the matrix M. The objective of a relaxation method is to choose the splitting matrix M such that the sequence is easily computed and the sequence converges rapidly to a solution.

Let A be split into $L + D + U$, where L is strictly lower triangular, D is a diagonal matrix, and U is strictly upper triangular. Note that these matrices are different from the L and U obtained from LU factorization. The vector x can then be solved for in an iterative manner using the Jacobi relaxation method,

$$x^{k+1} = -D^{-1}\left((L+U)\, x^k - b\right) \tag{2.46}$$

or identically in scalar form,

$$x_i^{k+1} = -\sum_{\substack{j=1 \\ j \neq i}}^{n} \left(\frac{a_{ij}}{a_{ii}}\right) x_j^k + \frac{b_i}{a_{ii}} \quad 1 \leq i \leq n,\ k \geq 0 \tag{2.47}$$

In the Jacobi relaxation method, all of the updates of the approximation vector x^{k+1} are obtained by using only the components of the previous approximation vector x^k. Therefore this method is also sometimes called the method of simultaneous displacements.

The Gauss-Seidel relaxation method is similar:

$$x^{k+1} = -\left(L + D\right)^{-1}\left(Ux^k - b\right) \tag{2.48}$$

or in scalar form

$$x_i^{k+1} = -\sum_{j=1}^{i-1} \left(\frac{a_{ij}}{a_{jj}}\right) x_j^{k+1} - \sum_{j=i+1}^{n} \left(\frac{a_{ij}}{a_{ii}}\right) x_j^k + \frac{b_i}{a_{ii}} \quad 1 \leq i \leq n,\ k \geq 0 \tag{2.49}$$

The Gauss-Seidel method has the advantage that each new update x_i^{k+1} relies only on previously computed values at that iteration: $x_1^{k+1}, x_2^{k+1}, \ldots, x_{i-1}^{k+1}$. Since the states are updated one-by-one, the new values can be stored in the same locations held by the old values, thus reducing the storage requirements.

Since relaxation methods are iterative, it is essential to determine under what conditions they are guaranteed to converge to the exact solution

$$x^* = A^{-1}b \tag{2.50}$$

It is well known that a necessary and sufficient condition for the Jacobi relaxation method to converge given any initial guess x_0 is that all eigenvalues of

$$M_J \triangleq -D^{-1}\left(L + U\right) \tag{2.51}$$

must lie within the unit circle in the complex plane [43]. Similarly, the eigenvalues of

$$M_{GS} \triangleq -\left(L + D\right)^{-1} U \tag{2.52}$$

must lie within the unit circle in the complex plane for the Gauss-Seidel relaxation algorithm to converge for any initial guess x_0. In practice, these conditions are difficult to confirm. There are several more general conditions that are easily confirmed under which convergence in guaranteed. In particular, if A is strictly diagonally dominant, then both the Jacobi and Gauss-Seidel methods are guaranteed to converge to the exact solution.

The initial vector x_0 can be arbitrary; however if a good guess of the solution is available it should be used for x_0 to produce more rapid convergence to within some pre-defined tolerance.

In general, the Gauss-Seidel method converges faster than the Jacobi for most classes of problems. If A is lower-triangular, the Gauss-Seidel method will converge in one iteration to the exact solution, whereas the Jacobi method will take n iterations. The Jacobi method has the advantage, however, that at each iteration, each x_i^{k+1} is independent of all other x_j^{k+1} for $j \neq i$. Thus the computation of all x_i^{k+1} can proceed in parallel. This method is therefore well suited to parallel processing [28].

Both the Jacobi and Gauss-Seidel methods can be generalized to the block-Jacobi and block-Gauss-Seidel methods where A is split into block matrices $L + D + U$, where D is block diagonal and L and U are lower- and upper-block triangular respectively. The same necessary and sufficient convergence conditions exist for the block case as for the scalar case, that is, the eigenvalues of M_J and M_{GS} must lie within the unit circle in the complex plane.

Example 2.5

Solve

$$
\begin{bmatrix}
-10 & 2 & 3 & 6 \\
0 & -9 & 1 & 4 \\
2 & 6 & -12 & 2 \\
3 & 1 & 0 & -8
\end{bmatrix}
x =
\begin{bmatrix}
1 \\
2 \\
3 \\
4
\end{bmatrix}
\tag{2.53}
$$

for x using (1) the Gauss-Seidel method, and (2) the Jacobi method.

Solution 2.5 The Gauss-Seidel method given in equation (2.49) with the initial vector $x = [0\ 0\ 0\ 0]$ leads to the following updates:

k	x_1	x_2	x_3	x_4
1	0.0000	0.0000	0.0000	0.0000
2	-0.1000	-0.2222	-0.3778	-0.5653
3	-0.5969	-0.5154	-0.7014	-0.7883
4	-0.8865	-0.6505	-0.8544	-0.9137
5	-1.0347	-0.7233	-0.9364	-0.9784
6	-1.1126	-0.7611	-0.9791	-1.0124
7	-1.1534	-0.7809	-1.0014	-1.0301
8	-1.1747	-0.7913	-1.0131	-1.0394
9	-1.1859	-0.7968	-1.0193	-1.0443
10	-1.1917	-0.7996	-1.0225	-1.0468
11	-1.1948	-0.8011	-1.0241	-1.0482
12	-1.1964	-0.8019	-1.0250	-1.0489
13	-1.1972	-0.8023	-1.0255	-1.0492
14	-1.1976	-0.8025	-1.0257	-1.0494
15	-1.1979	-0.8026	-1.0259	-1.0495
16	-1.1980	-0.8027	-1.0259	-1.0496

The Gauss-Seidel iterates have converged to the solution

$$x = [-1.1980 \ -0.8027 \ -1.0259 \ -1.0496]^T$$

From equation (2.47) and using the initial vector $x = [0\ 0\ 0\ 0]$, the following updates are obtained for the Jacobi method:

k	x_1	x_2	x_3	x_4
1	0.0000	0.0000	0.0000	0.0000
2	-0.1000	-0.2222	-0.2500	-0.5000
3	-0.5194	-0.4722	-0.4611	-0.5653
4	-0.6719	-0.5247	-0.6669	-0.7538
5	-0.8573	-0.6314	-0.7500	-0.8176
6	-0.9418	-0.6689	-0.8448	-0.9004
7	-1.0275	-0.7163	-0.8915	-0.9368
8	-1.0728	-0.7376	-0.9355	-0.9748
9	-1.1131	-0.7594	-0.9601	-0.9945
10	-1.1366	-0.7709	-0.9810	-1.0123
11	-1.1559	-0.7811	-0.9936	-1.0226
12	-1.1679	-0.7871	-1.0037	-1.0311
13	-1.1772	-0.7920	-1.0100	-1.0363
14	-1.1832	-0.7950	-1.0149	-1.0404
15	-1.1877	-0.7974	-1.0181	-1.0431
16	-1.1908	-0.7989	-1.0205	-1.0451
17	-1.1930	-0.8001	-1.0221	-1.0464
18	-1.1945	-0.8009	-1.0233	-1.0474
19	-1.1956	-0.8014	-1.0241	-1.0480
20	-1.1963	-0.8018	-1.0247	-1.0485
21	-1.1969	-0.8021	-1.0250	-1.0489
22	-1.1972	-0.8023	-1.0253	-1.0491
23	-1.1975	-0.8024	-1.0255	-1.0492
24	-1.1977	-0.8025	-1.0257	-1.0494
25	-1.1978	-0.8026	-1.0258	-1.0494

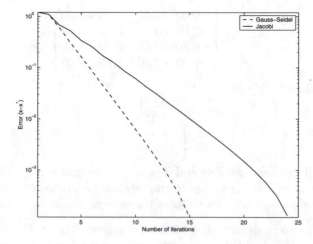

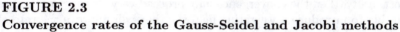

FIGURE 2.3
Convergence rates of the Gauss-Seidel and Jacobi methods

The Jacobi iterates have converged to the same solution as the Gauss-Seidel method. The error in the iterates is shown in Figure 2.3 on a semi-log scale, where the error is defined as the maximum $|(x_i^k - x_i^*)|$ for all $i = 1, \ldots, 4$. Both the Gauss-Seidel and the Jacobi methods exhibit *linear convergence*, but the Gauss-Seidel converges with a steeper slope and will therefore reach the convergence tolerance sooner for the same initial condition. ■

Example 2.6
Repeat Example 2.2 using the Jacobi iterative method.

Solution 2.6 Repeating the solution procedure of Example 2.5 yields the following iterations for the Jacobi method:

k	x_1	x_2	x_3	x_4
1	0	0	0	0
2	1.0000	1.0000	0.2000	0.2500
3	-4.8000	-2.1500	-1.6000	-2.8500
4	36.6500	22.3500	9.8900	14.9250
5	-225.0100	-136.8550	-66.4100	-110.6950

Obviously these iterates are not converging. To understand why they are diverging, consider the iterative matrix for the Jacobi matrix:

$$M_J = -D^{-1}(L + U)$$

$$= \begin{bmatrix} 0.00 & -3.00 & -4.00 & -8.00 \\ -2.00 & 0.00 & -2.00 & -3.00 \\ -0.80 & -0.60 & 0.00 & -1.60 \\ -2.25 & -0.50 & -1.75 & 0.00 \end{bmatrix}$$

The eigenvalues of M_J are

$$\begin{bmatrix} -6.6212 \\ 4.3574 \\ 1.2072 \\ 1.0566 \end{bmatrix}$$

which are all greater than one and lie outside the unit circle. Therefore, the Jacobi method will not converge to the solution regardless of choice of initial condition and cannot be used to solve the system of Example 2.2. ∎

If the largest eigenvalue of the iterative matrix M_J or M_{GS} is less than, but almost, unity, then the convergence may proceed very slowly. In this case it is desirable to introduce a weighting factor ω that will improve the rate of convergence. From

$$x^{k+1} = -(L+D)^{-1}(Ux^k - b) \tag{2.54}$$

it follows that

$$x^{k+1} = x^k - D^{-1}(Lx^{k+1} + (D+U)x^k - b) \tag{2.55}$$

A new iterative method can be defined with the weighting factor ω such that

$$x^{k+1} = x^k - \omega D^{-1}(Lx^{k+1} + (D+U)x^k - b) \tag{2.56}$$

This method is known as the *successive overrelaxation (SOR)* method with relaxation coefficient $\omega > 0$. Note that if the relaxation iterates converge, they converge to the solution $x^* = A^{-1}b$. One necessary condition for the SOR method to be convergent is that $0 < \omega < 2$ [19]. The calculation of the optimal value for ω is difficult, except in a few simple cases. The optimal value is usually determined through trial and error, but analysis shows that for systems larger than $n = 30$, the optimal SOR can be more than forty times faster than the Jacobi method [19]. The improvement on the speed of convergence often improves as n increases.

2.5   Conjugate Gradient Methods

Another common iterative method for solving $Ax = b$ is the *conjugate gradient* method. This method can be considered a minimization method for the function

$$E(x) = \|Ax - b\|^2 \tag{2.57}$$

along a succession of rays. One attractive feature of this method is that it is guaranteed to converge in at most n steps (neglecting round-off error) if the A matrix is positive definite. The conjugate gradient method is most frequently used instead of Gaussian elimination if the A matrix is very large and sparse, in which case the solution may be obtained in less than n steps. This is especially true if the A matrix is well conditioned. If the matrix is ill conditioned, then round-off errors may prevent the algorithm from obtaining a sufficiently accurate solution after n steps.

As the conjugate gradient method progresses, each error function is associated with a specific ray, or orthogonal expansion. Therefore the conjugate gradient method is reduced to the process of generating the orthogonal vectors and finding the proper coefficients to represent the desired solution.

All iterative methods for solving $Ax = b$ define an iterative process such that

$$x^{k+1} = x^k + \alpha_{k+1}\rho_{k+1} \tag{2.58}$$

where x^{k+1} is the updated value, α_k is the steplength, and ρ_k defines the direction $\in R^n$ in which the algorithm moves to update the estimate.

Let the residual, or mismatch, vector at step k be given by

$$r_k = Ax_k - b \tag{2.59}$$

and the error function given by

$$E_k(x_k) = \|Ax_k - b\|^2 \tag{2.60}$$

Then the coefficient that minimizes the error function at step $k+1$ is

$$\alpha_{k+1} = \frac{\|A^T r_k\|^2}{\|A\rho_{k+1}\|^2} \tag{2.61}$$

This has the geometric interpretation of minimizing E_{k+1} along the ray defined by ρ_{k+1}. Further, an improved algorithm is one that seeks the minimum of E_{k+1} in a plane spanned by two direction vectors, such that

$$x^{k+1} = x^k + \alpha_{k+1}(\rho_{k+1} + \beta_{k+1}\sigma_{k+1}) \tag{2.62}$$

where the rays ρ_{k+1} and σ_{k+1} span a plane in R^n. The process of selecting direction vectors and coefficients to minimize the error function E_{k+1} is optimized when the chosen vectors are orthogonal, such that

$$\langle A\rho_{k+1}, A\sigma_{k+1} \rangle = 0 \tag{2.63}$$

where $\langle \cdot \rangle$ denotes inner product. Vectors that satisfy the orthogonality condition of equation (2.63) are said to be mutually conjugate with respect to the operator $A^T A$, where A^T is the conjugate transpose of A. One method of choosing appropriate vectors is to choose σ_{k+1} as a vector orthogonal to

ρ_k, thus eliminating the need to specify two orthogonal vectors at each step. While this simplifies the procedure, there is now an implicit recursive dependence for generating the ρ vectors.

Conjugate Gradient Algorithm for Solving $Ax = b$

Initialization: Let $k = 1$, and

$$r_0 = Ax^0 - b \tag{2.64}$$
$$\rho_0 = -A^T r_0 \tag{2.65}$$

While $\|r_k\| \geq \varepsilon$

$$\alpha_{k+1} = \frac{\|A^T r_k\|^2}{\|A\rho_k\|^2} \tag{2.66}$$

$$x^{k+1} = x^k + \alpha_{k+1}\rho_k \tag{2.67}$$

$$r_{k+1} = Ax^{k+1} - b \tag{2.68}$$

$$B_{k+1} = \frac{\|A^T r_{k+1}\|^2}{\|A^T r_k\|^2} \tag{2.69}$$

$$\rho_{k+1} = -A^T r_{k+1} + B_{k+1}\rho_k \tag{2.70}$$

$$k = k + 1 \tag{2.71}$$

For an arbitrary nonsingular positive definite matrix A, the conjugate gradient method will produce a solution in at most n steps (neglecting round-off error). This is a direct consequence of the fact that the n direction vectors $\rho_0, \rho_1, \ldots$ span the solution space. Finite step termination is a significant advantage of the conjugate gradient method over other iterative methods such as relaxation methods.

Example 2.7

Repeat Example 2.5 using the conjugate gradient method.

Solution 2.7 The problem of Example 2.5 is repeated here for convenience: Solve

$$\begin{bmatrix} -10 & 2 & 3 & 6 \\ 0 & -9 & 1 & 4 \\ 2 & 6 & -12 & 2 \\ 3 & 1 & 0 & -8 \end{bmatrix} x = \begin{bmatrix} 1 \\ 2 \\ 3 \\ 4 \end{bmatrix} \tag{2.72}$$

with $x^0 = [0\ 0\ 0\ 0]^T$.

Initialization: Let $k = 1$, and

$$r_0 = Ax^0 - b = \begin{bmatrix} -1 \\ -2 \\ -3 \\ -4 \end{bmatrix} \qquad (2.73)$$

$$\rho_0 = -A^T r_0 = \begin{bmatrix} 8 \\ 6 \\ -31 \\ -12 \end{bmatrix} \qquad (2.74)$$

The initial error is

$$E_0 = \|r_0\| = 5.4772 \qquad (2.75)$$

Iteration 1

$$\alpha_0 = \frac{\|A^T r_0\|^2}{\|A\rho_0\|^2} = 0.0049 \qquad (2.76)$$

$$x^1 = x^0 + \alpha_0 \rho_0 = \begin{bmatrix} 0.0389 \\ 0.0292 \\ -0.1507 \\ -0.0583 \end{bmatrix} \qquad (2.77)$$

$$r_1 = Ax^1 - b = \begin{bmatrix} -2.1328 \\ -2.6466 \\ -1.0553 \\ -3.3874 \end{bmatrix} \qquad (2.78)$$

$$B_1 = \frac{\|A^T r_1\|^2}{\|A^T r_0\|^2} = 0.1613 \qquad (2.79)$$

$$\rho_1 = -A^T r_1 + B_1 \rho_0 = \begin{bmatrix} -7.7644 \\ -8.8668 \\ -8.6195 \\ -3.5414 \end{bmatrix} \qquad (2.80)$$

and the error is

$$E_1 = \|r_1\|_2 = 4.9134 \qquad (2.81)$$

Similarly for iterations 2–4:

k	α_k	x^k	r_k	B_k	ρ_k	$\|r_k\|$
2	0.0464	$\begin{bmatrix} -0.3211 \\ -0.3820 \\ -0.5504 \\ -0.2225 \end{bmatrix}$	$\begin{bmatrix} -1.5391 \\ -0.0030 \\ 0.2254 \\ -3.5649 \end{bmatrix}$	2.5562	$\begin{bmatrix} -24.9948 \\ -17.4017 \\ -14.7079 \\ -28.7765 \end{bmatrix}$	3.8895
3	0.0237	$\begin{bmatrix} -0.9133 \\ -0.7942 \\ -0.8988 \\ -0.9043 \end{bmatrix}$	$\begin{bmatrix} -1.5779 \\ -0.6320 \\ -0.6146 \\ -0.2998 \end{bmatrix}$	0.7949	$\begin{bmatrix} -33.5195 \\ -1.0021 \\ -14.9648 \\ -17.1040 \end{bmatrix}$	1.8322
4	0.0085	$\begin{bmatrix} -1.1981 \\ -0.8027 \\ -1.0260 \\ -1.0496 \end{bmatrix}$	$\begin{bmatrix} 0.0000 \\ 0.0000 \\ 0.0000 \\ 0.0000 \end{bmatrix}$	0.0000	$\begin{bmatrix} 0.0000 \\ 0.0000 \\ 0.0000 \\ 0.0000 \end{bmatrix}$	0.0000

The iterations converged in four steps as the algorithm guarantees. ∎

Unfortunately for a general linear system, the conjugate gradient method requires significantly more multiplications and divisions than does the LU factorization method. The conjugate gradient method is more numerically competitive for matrices that are very large and sparse or that have a special structure that cannot be easily be handled by LU factorization. In some cases, the speed of convergence of the conjugate gradient method can be improved by *preconditioning*.

As seen with the Gauss-Seidel and Jacobi iteration, the convergence rate of iterative algorithms is closely related to the eigenvalue spectrum of the iterative matrix. Consequently, scaling or matrix transformation that converts the original system of equations into one with a better eigenvalue spectrum might significantly improve the rate of convergence. This procedure is known as preconditioning. A number of systematic approaches for sparse-matrix preconditioning have been developed in which the basic approach is to convert the system

$$Ax = b$$

into an equivalent system

$$M^{-1}Ax = M^{-1}b$$

where M^{-1} approximates A^{-1}. For example, one such approach might be to use an incomplete LU factorization, where the LU factorization method is applied but all fills are neglected. If the A matrix is diagonally dominant, a simple approximation to M^{-1} is

$$M^{-1} = \begin{bmatrix} \frac{1}{A(1,1)} & & & \\ & \frac{1}{A(2,2)} & & \\ & & \ddots & \\ & & & \frac{1}{A(n,n)} \end{bmatrix} \tag{2.82}$$

This preconditioning strategy scales the system of equations so that the entries along the main diagonal are all equal. This procedure can compensate for orders-of-magnitude differences in scale. Note that scaling will not have any effect on matrices that are inherently ill conditioned.

Example 2.8
Repeat Example 2.7 using a preconditioner.

Solution 2.8 Let M^{-1} be defined as in equation (2.82). Thus

$$M^{-1} = \begin{bmatrix} -\frac{1}{10} & & & \\ & -\frac{1}{9} & & \\ & & \frac{1}{12} & \\ & & & -\frac{1}{8} \end{bmatrix} \tag{2.83}$$

$$A' = M^{-1}A = \begin{bmatrix} 1.0000 & -0.2000 & -0.3000 & -0.6000 \\ 0.0000 & 1.0000 & -0.1111 & -0.4444 \\ -0.1667 & -0.5000 & 1.0000 & -0.1667 \\ -0.3750 & -0.1250 & 0.0000 & 1.0000 \end{bmatrix} \tag{2.84}$$

$$b' = M^{-1}b = \begin{bmatrix} -0.1000 \\ -0.2222 \\ -0.2500 \\ -0.5000 \end{bmatrix} \tag{2.85}$$

Solving $A'x = b'$ using the conjugate gradient method yields the following set of errors:

k	E_k
0	0.6098
1	0.5500
2	0.3559
3	0.1131
4	0.0000

Although it takes the same number of iterations to converge, note that the errors for $k < 4$ are much smaller than in Example 2.7. For large systems, it is conceivable that the error would be decreased sufficiently rapidly to terminate the iterations prior to the n-th step. This method is also useful if only an approximate solution to x is desired. ∎

2.6 Problems

1. Show that the number of multiplications and divisions required in the LU factorization of an $n \times n$ square matrix is $n(n^2 - 1)/3$.

2. Consider the system $Ax = b$, where

$$a_{ij} = \frac{1}{i+j-1} \quad i,j = 1,\ldots,4$$

and

$$b_i = \frac{1}{3} \sum_{j=1}^{4} a_{ij}$$

Using only four decimal places of accuracy, solve this system using LU factorization

   (a) no pivoting
   (b) partial pivoting

Comment on the differences in solutions (if any).

3. Prove that the matrix

$$A = \begin{bmatrix} 0 & 1 \\ 1 & 1 \end{bmatrix}$$

does not have an LU factorization.

4. Assuming that an LU factorization of A is available, write an algorithm to solve the equation $x^T A = b^T$.

5. For the following matrix, find $A = LU$ and $PA = LU$ using partial pivoting

$$A = \begin{bmatrix} 6 & -2 & 2 & 4 \\ 12 & -8 & 4 & 10 \\ 3 & -13 & 3 & 3 \\ -6 & 4 & 2 & -18 \end{bmatrix}$$

6. Write an LU factorization-based algorithm to find the inverse of any nonsingular matrix A.

7. Consider an $n \times n$ tridiagonal matrix of the form

$$T_a = \begin{bmatrix} a & -1 & & & & \\ -1 & a & -1 & & & \\ & -1 & a & -1 & & \\ & & -1 & a & -1 & \\ & & & -1 & a & -1 \\ & & & & -1 & a \end{bmatrix}$$

where a is a real number.

(a) Verify that the eigenvalues of T_a are given by

$$\lambda_j = a - 2\cos(j\theta) \quad j = 1, \ldots, n$$

where

$$\theta = \frac{\pi}{n+1}$$

(b) Let $a = 2$.

i. Will the Jacobi iteration converge for this matrix?

ii. Will the Gauss-Seidel iteration converge for this matrix?

8. Apply the Gauss-Seidel iteration to the system

$$A = \begin{bmatrix} 0.96326 & 0.81321 \\ 0.81321 & 0.68654 \end{bmatrix}$$

$$b = \begin{bmatrix} 0.88824 \\ 0.74988 \end{bmatrix}$$

Use $x^0 = [0.33116 \ 0.70000]^T$ and explain what happens.

9. Solve the matrix of problem 2 using the conjugate gradient method.

10. An alternative conjugate gradient algorithm for solving $Ax = b$ may be based on the error functional $E_k(x^k) = \langle x^k - x, x^k - x \rangle$ where $\langle \cdot \rangle$ denotes inner product. The solution is given as

$$x^{k+1} = x^k + \alpha_k \sigma_k$$

Using $\sigma_1 = -A^T r_0$ and $\sigma_{k+1} = -A^T r_k + \beta_k \sigma_k$, derive this conjugate gradient algorithm. The coefficients α_k and β_k can be expressed as

$$\alpha_{k+1} = \frac{\|r_k\|^2}{\|\sigma_{k+1}\|^2}$$

$$\beta_{k+1} = \frac{\|r_{k+1}\|^2}{\|r_k\|^2}$$

Repeat Example 2.7 using this conjugate gradient algorithm.

11. Write a computer program (as a set of subroutines) that will generate for any non-singular matrix A, a permutation matrix P, and a matrix Q such that

$$PA = LU$$

where

$$L = \begin{bmatrix} l_{11} & 0 & 0 & \cdots & 0 \\ l_{21} & l_{22} & 0 & \cdots & 0 \\ l_{31} & l_{32} & l_{33} & \cdots & 0 \\ \vdots & \vdots & \vdots & \vdots & \vdots \\ l_{n1} & l_{n2} & l_{n3} & \cdots & l_{nn} \end{bmatrix} \quad \text{and} \quad U = \begin{bmatrix} 1 & u_{12} & u_{13} & \cdots & u_{1n} \\ 0 & 1 & u_{23} & \cdots & u_{2n} \\ 0 & 0 & 1 & \cdots & u_{3n} \\ \vdots & \vdots & \vdots & \vdots & \vdots \\ 0 & 0 & 0 & \cdots & 1 \end{bmatrix}$$

and

$$Q = L + U - I$$

Your program should define the subroutines *lufact* and *permute*, where *permute* is embedded in *lufact*.

12. For the following non-singular matrices, use the subroutines of Problem 11 and obtain matrices P and Q in each of the following cases:

(a)

$$\begin{bmatrix} 0 & 0 & 1 \\ 3 & 1 & 4 \\ 2 & 1 & 0 \end{bmatrix}$$

(b)

$$\begin{bmatrix} 10^{-10} & 0 & 0 & 1 \\ 0 & 0 & 1 & 4 \\ 0 & 2 & 1 & 0 \\ 1 & 0 & 0 & 0 \end{bmatrix}$$

13. Write a forward and backward substitution subroutine called *sub* which accepts as an input the LU factors and the b vector, and outputs the solution vector x to the linear system of equations $Ax = b$.

14. Using the subroutines of Problems 11 and 13, solve the following system of equations

$$\begin{bmatrix} 2 & 5 & 6 & 11 \\ 4 & 6 & 8 & 2 \\ 4 & 3 & 7 & 0 \\ 1 & 26 & 3 & 4 \end{bmatrix} \begin{bmatrix} x_1 \\ x_2 \\ x_3 \\ x_4 \end{bmatrix} = \begin{bmatrix} 1 \\ 1 \\ 1 \\ 1 \end{bmatrix}$$

Chapter 3

Systems of Nonlinear Equations

Many systems can be modeled generically as

$$F(x) = 0 \qquad (3.1)$$

where x is an n-vector and F represents a nonlinear mapping with both its domain and range in the n-dimensional real linear space R^n. The mapping F can also be interpreted as being an n-vector of functions

$$F(x) = \begin{bmatrix} f_1(x_1, x_2, \ldots, x_n) \\ f_2(x_1, x_2, \ldots, x_n) \\ \vdots \\ f_n(x_1, x_2, \ldots, x_n) \end{bmatrix} = 0 \qquad (3.2)$$

where at least one of the functions is nonlinear. Each function may or may not involve all n states x_i, but it is assumed that every state appears at least once in the set of functions. The solution x^* of the nonlinear system cannot, in general, be expressed in closed form. Thus nonlinear systems are usually solved numerically. In many cases, it is possible to find an approximate solution $\hat{x}$ arbitrarily close to the actual solution x^*, by replacing each approximation with successively better (more accurate) approximations until

$$F(\hat{x}) \approx 0.$$

Such methods are usually *iterative*. An iterative solution is one in which an initial guess (x^0) to the solution is used to create a sequence x^0, x^1, x^2, ... that (hopefully) converges arbitrarily close to the desired solution x^*.

Three principal issues arise with the use of iterative methods, namely

1. Is the iterative process well defined? That is, can it be successively applied without numerical difficulties?

2. Do the iterates (i.e., the sequence of updates) converge to a solution of equation (3.1)? Is the solution the desired solution?

3. How economical is the entire solution process?

The complete (or partial) answers to these issues are enough to fill several volumes, and as such cannot be discussed in complete detail in this chapter.

These issues, however, are central to the solution of nonlinear systems and cannot be fully ignored. Therefore this chapter will endeavor to provide sufficient detail for the reader to be aware of the advantages (and disadvantages) of different types of iterative methods without providing exhaustive coverage.

3.1   Fixed Point Iteration

Solving a system of nonlinear equations is a complex problem. To better understand the mechanisms involved in a large-scale system, it is instructive to first consider the one dimensional, or scalar, nonlinear system

$$f(x) = 0. \tag{3.3}$$

One approach to solving any nonlinear equation is the tried-and-true "trial and error" method that most engineering and science students have used at one time or another in their careers.

Example 3.1
Find the solution to

$$f(x) = x^2 - 5x + 4 = 0 \tag{3.4}$$

Solution 3.1 This is a quadratic equation that has a closed form solution. The two solutions are

$$x_1^*, x_2^* = \frac{5 \pm \sqrt{(-5)^2 - 4 \cdot 4}}{2} = 1, \ 4$$

If a closed form solution did not exist, however, one approach would be to use a trial and error approach. Since the solution occurs when $f(x) = 0$, the value of $f(x)$ can be monitored and used to refine the estimates to x^*.

k	x	$f(x)$
0	0	$0 - 0 + 4 = 4 > 0$
1	2	$4 - 10 + 4 = -2 < 0$
2	0.5	$0.25 - 2.5 + 4 = 1.75 > 0$
3	1.5	$2.25 - 7.5 + 4 = -1.25 < 0$

By noting the sign of the function and whether or not it changes sign, the interval in which the solution lies can be successively narrowed. If a function $f(x)$ is continuous and $f(a) \cdot f(b) < 0$, then the equation $f(x) = 0$ has at least one solution in the interval (a, b). Since $f(0.5) > 0$ and $f(1.5) < 0$ it can be concluded that one of the solutions lies in the interval $(0.5, 1.5)$. ∎

This process, however, tends to be tedious and there is no guidance to determine what the next guess should be other than the bounds established by the change in sign of $f(x)$. A better method would be to write the sequence of updates in terms of the previous guesses. Thus, an iterative function can be defined as:

$$I: \quad x^{k+1} = g\left(x^k\right), \quad k = 1, \ldots, \infty \tag{3.5}$$

This is known as a fixed-point iteration because at the solution

$$x^* = g\left(x^*\right) \tag{3.6}$$

Example 3.2
Find the solution to equation (3.4) using a fixed point iteration.

Solution 3.2 Equation (3.4) can be rewritten as

$$x = \frac{x^2 + 4}{5} \tag{3.7}$$

Adopting the notation of equation (3.5), the iterative function becomes

$$x^{k+1} = g\left(x^k\right) = \frac{\left(x^k\right)^2 + 4}{5} \tag{3.8}$$

Using this iterative function, the estimates to x^* are:

k	x^k	$g\left(x^k\right)$
0	0	$\frac{0+4}{5} = 0.8$
1	0.8	$\frac{0.64+4}{5} = 0.928$
2	0.928	$\frac{0.856+4}{5} = 0.971$
3	0.971	$\frac{0.943+4}{5} = 0.989$

It is obvious that this sequence is converging to the solution $x^* = 1$.
Now consider the same example, except with a different initial guess:

k	x^k	$g\left(x^k\right)$
0	5	$\frac{25+4}{5} = 5.8$
1	5.8	$\frac{33.64+4}{5} = 7.528$
2	7.528	$\frac{56.67+4}{5} = 12.134$

In this case, the iterates are increasing rapidly and after a few more iterations would approach infinity. In this case, it is said that the iteration is *diverging*. ∎

This example brings up two very important points: will a sequence of iterates converge and, if so, to what solution will they converge? In order to address these questions, consider first a graphical interpretation of Example 3.2. Plotting both sides of the function in equation (3.7) yields the two lines shown in Figure 3.1. These two lines intersect at the same two points in which

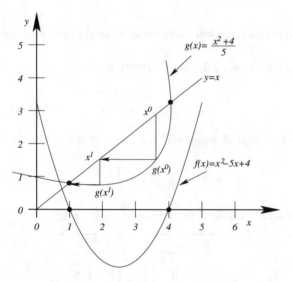

FIGURE 3.1
Graphical interpretation of the fixed point iteration

the original function $f(x) = 0$. The fixed point iteration works by finding this intersection. Consider the initial guess x^0 shown in Figure 3.1. The function $g(x)$ evaluated at x^0 gives the updated iterate x^1. Thus a vertical line projected from x^0 points to $g(x^0)$ and a horizontal line projected from $g(x^0)$ gives x^1.

The projection of the function $g(x^1)$ yields x^2. Similar vertical and horizontal projections will eventually lead directly to the point at which the two lines intersect. In this way, the solution to the original function $f(x)$ can be obtained.

In this example, the solution $x^* = 1$ is the *point of attraction* of the fixed point iteration. A point x^* is said to be a point of attraction of an iterative function I if there exists an open neighborhood S_0 of x^* such that for all initial guesses x^0 in the subset S_0 of S, the iterates will remain in S and

$$\lim_{k \to \infty} x^k = x^* \tag{3.9}$$

The neighborhood S_0 is called the *domain of attraction* of x^* [26]. This concept is illustrated in Figure 3.2 and implies that the iterates of I will converge to x^* whenever x^0 is sufficiently close to x^*. In Example 3.2, the fixed point $x^* = 1$ is a point of attraction of

$$I: \quad x^{k+1} = \frac{\left(x^k\right)^2 + 4}{5}$$

whereas $x^* = 4$ is not. The domain of attraction of $x^* = 1$ is all x in the domain $-\infty < x < 4$.

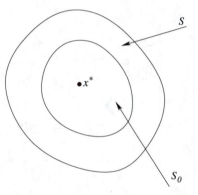

FIGURE 3.2
Domain of attraction of x^*

One method of determining whether or not an iterative process will converge is given by the **Ostrowski** theorem [26]. This theorem states that if the iterative process

$$I: \quad x^{k+1} = g\left(x^k\right), \quad k = 1, \dots, \infty$$

has a fixed point x^* and is continuous and differentiable at x^*, and if $\left|\frac{\partial g(x^*)}{\partial x}\right| <$ 1, then x^* is a point of attraction of I.

Example 3.3
Determine whether $x^* = 1$ and $x^* = 4$ are points of attraction of the iterative function of equation (3.8).

Solution 3.3 The derivative of the iterative process I in equation (3.8) is

$$\left|\frac{\partial g\left(x\right)}{\partial x}\right| = \left|\frac{2}{5}x\right|$$

Thus, for $x^* = 1$, $\left|\frac{2}{5}x^*\right| = \frac{2}{5} < 1$ and $x^* = 1$ is a point of attraction of I. For $x^* = 4$: $\left|\frac{2}{5}x^*\right| = \frac{2}{5}\left(4\right) = \frac{8}{5} > 1$; thus, $x^* = 4$ is not a point of attraction of I. ∎

In many cases however, it is difficult to determine whether or not an iteration will converge since the value of x^* is not known ahead of time. In some cases, a series of iterates will appear to be converging, but will not approach x^* even as $k \to \infty$. Consider the fixed point iteration shown in Figure 3.3. The iterates will initially approach the fixed point x^*, but once they reach a certain point, they will no longer converge, but will continue to oscillate about the fixed point as indicated by the dashed arrows.

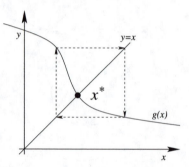

FIGURE 3.3
Non-converging iteration

3.2   Newton-Raphson Iteration

Several iterative methods offer more robust convergence behavior than the simple fixed point iteration described in the previous section. One of the most widely used iterative methods is the *Newton-Raphson* iterative method. This method can also be described by the iterative process

$$I: \quad x^{k+1} = g\left(x^k\right), \quad k = 1,\ldots,\infty$$

but frequently offers better convergence properties than the fixed point iteration.

Consider again the scalar nonlinear function

$$f(x^*) = 0 \tag{3.10}$$

Expanding this function in a Taylor's series expansion about the point x^k yields

$$f(x^*) = f\left(x^k\right) + \left.\frac{\partial f}{\partial x}\right|_{x^k} \left(x^* - x^k\right) + \frac{1}{2!}\left.\frac{\partial^2 f}{\partial x^2}\right|_{x^k} \left(x^* - x^k\right)^2 + \ldots = 0 \tag{3.11}$$

If it is assumed that the iterates will converge to x^* as $k \to \infty$, then the updated guess x^{k+1} can be substituted for x^*, yielding

$$f(x^{k+1}) = f\left(x^k\right) + \left.\frac{\partial f}{\partial x}\right|_{x^k} \left(x^{k+1} - x^k\right) + \frac{1}{2!}\left.\frac{\partial^2 f}{\partial x^2}\right|_{x^k} \left(x^{k+1} - x^k\right)^2 + \ldots = 0$$

$$\tag{3.12}$$

If the initial guess is "sufficiently close" to x^* and within the domain of attraction of x^*, then the higher order terms of the expansion can be neglected, yielding

$$f(x^{k+1}) = f\left(x^k\right) + \left.\frac{\partial f}{\partial x}\right|_{x^k} \left(x^{k+1} - x^k\right) \approx 0 \tag{3.13}$$

Solving directly for x^{k+1} as a function of x^k yields the following iterative function:

$$I: \quad x^{k+1} = x^k - \left[\left.\frac{\partial f}{\partial x}\right|_{x^k}\right]^{-1} f\left(x^k\right) \tag{3.14}$$

which is the well-known Newton-Raphson iterative method.

The Newton-Raphson method also lends itself to a graphical interpretation. Consider the same function as in Example 3.2 plotted in Figure 3.4. In this method, the slope of the function evaluated at the current iteration is used to produce the next guess. For any guess x^k, there corresponds a point on the function $f\left(x^k\right)$ with slope

$$\left.\frac{\partial f}{\partial x}\right|_{x=x^k}$$

Therefore, the next guess x^{k+1} is simply the intersection of the slope and the x-axis. This process is repeated until the guesses are sufficiently close to the solution x^*. An iteration is said to have converged at x^k if

$$\left|f\left(x^k\right)\right| < \varepsilon$$

where ε is some pre-determined tolerance.

Example 3.4
Repeat Example 3.2 using a Newton-Raphson iteration.

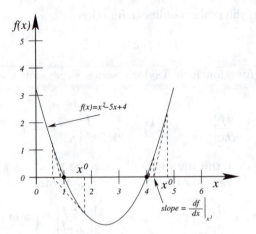

FIGURE 3.4
Graphical interpretation of the Newton-Raphson method

Solution 3.4 Using the Newton-Raphson method of equation (3.14), the iterative function is given by:

$$I: \quad x^{k+1} = x^k - \frac{\left(x^k\right)^2 - 5x^k + 4}{2x^k - 5} \tag{3.15}$$

Using this iterative function, the estimates to x^* from an initial guess of $x^0 = 3$ are:

k	x^k	$g\left(x^k\right)$
0	3	$3 - \frac{9-15+4}{6-5} = 5$
1	5	$5 - \frac{25-25+4}{10-5} = 4.2$
2	4.2	$4.2 - \frac{17.64-21+4}{8.4-5} = 4.012$

Similarly, the estimates to x^* from an initial guess of $x^0 = 2$ are:

k	x^k	$g\left(x^k\right)$
0	2	$2 - \frac{4-10+4}{4-5} = 0$
1	0	$0 - \frac{0-0+4}{0-5} = 0.8$
2	0.8	$0.8 - \frac{0.64-4+4}{1.6-5} = 0.988$

In this case, both solutions are points of attraction of the Newton-Raphson iteration. ■

In some cases however, the Newton-Raphson method will also fail to converge. Consider the function shown in Figure 3.5. An initial guess of x_a^0 will converge to the solution x_a^*. An initial guess of x_b^0 will converge to the solution x_b^*. However, an initial guess of x_c^0 will cause the iterates to get locked-in and oscillate in the region denoted by the dashed box without ever converging to a solution. This figure supports the assertion that if the initial guess is too far away from the actual solution, the iterates may not converge. Or conversely, the initial guess must be sufficiently close to the actual solution for the Newton-Raphson iteration to converge. This supports the initial assumption used to derive the Newton-Raphson algorithm in that *if the iterates were sufficiently close to the actual solution,* the higher-order terms of the Taylor series expansion could be neglected. If the iterates are not sufficiently close to the actual solution, these higher-order terms are significant and the assumption upon which the Newton-Raphson algorithm is based is not valid.

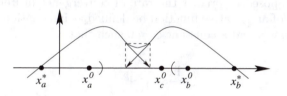

FIGURE 3.5
Newton-Raphson regions of convergence

3.2.1 Convergence Properties

Note that the rate of convergence to the solution in Example 3.4 is much faster than in Example 3.2. This is because the Newton-Raphson method exhibits *quadratic convergence*, whereas the fixed-point iteration exhibits only *linear convergence*. Linear convergence implies that once the iterates x^k are sufficiently close to the actual solution x^*, then the error

$$\varepsilon^k = \left| x^k - x^* \right| \tag{3.16}$$

will approach zero in a linear fashion. The convergence of Examples 3.2 and 3.4 is shown in Figure 3.6. Plotted on a log-scale plot, the error for the fixed point iteration is clearly linear, whereas the Newton-Raphson error exhibits

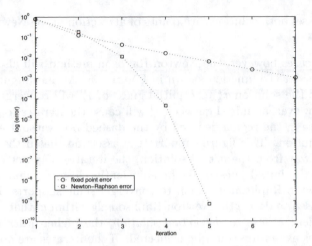

FIGURE 3.6
Non-converging iteration (fixed point vs. Newton-Raphson)

quadratic convergence until it becomes too small to plot. Numerous methods have been proposed to predict the rate of convergence of iterative methods. Let the error of an iterative function be defined as in equation (3.16). If there exists a number p and a constant $C \neq 0$ such that

$$\lim_{k \to \infty} \frac{|\varepsilon^{k+1}|}{|\varepsilon^k|^p} = C \tag{3.17}$$

then p is called the *order of convergence* of the iterative sequence and C is the *asympototic error constant*. If $p = 1$, the convergence is said to be *linear*. If $p = 2$, the convergence is *quadratic*, and if $p = 3$, the order of convergence is *cubic*. The Newton-Raphson method satisfies equation (3.17) with $p = 2$ if

$$C = \frac{1}{2} \frac{\left| \frac{d^2 f(x^*)}{dx^2} \right|}{\left| \frac{df(x^*)}{dx} \right|}$$

where $C \neq 0$ only if $\frac{d^2 f(x^*)}{dx^2} \neq 0$. Thus, for most functions, the Newton-Raphson method exhibits quadratic convergence.

3.2.2 The Newton-Raphson for Systems of Nonlinear Equations

In science and engineering, many applications give rise to *systems* of equations such as those in equation (3.2). With a few modifications, the Newton-Raphson method developed in the previous section can be extended to systems of nonlinear equations. Systems of equations can similarly be represented by

Taylor series expansions. By making the assumption once again that the initial guess is sufficiently close to the exact solution, then the multi-dimensional higher order terms can be neglected, yielding the Newton-Raphson method for n-dimensional systems:

$$x^{k+1} = x^k - \left[J\left(x^k \right) \right]^{-1} F\left(x^k \right) \tag{3.18}$$

where

$$x = \begin{bmatrix} x_1 \\ x_2 \\ x_3 \\ \vdots \\ x_n \end{bmatrix}$$

$$F\left(x^k \right) = \begin{bmatrix} f_1\left(x^k \right) \\ f_2\left(x^k \right) \\ f_3\left(x^k \right) \\ \vdots \\ f_n\left(x^k \right) \end{bmatrix}$$

and the Jacobian matrix $\left[J\left(x^k \right) \right]$ is given by

$$\left[J\left(x^k \right) \right] = \begin{bmatrix} \frac{\partial f_1}{\partial x_1} & \frac{\partial f_1}{\partial x_2} & \frac{\partial f_1}{\partial x_3} & \cdots & \frac{\partial f_1}{\partial x_n} \\[2mm] \frac{\partial f_2}{\partial x_1} & \frac{\partial f_2}{\partial x_2} & \frac{\partial f_2}{\partial x_3} & \cdots & \frac{\partial f_2}{\partial x_n} \\[2mm] \frac{\partial f_3}{\partial x_1} & \frac{\partial f_3}{\partial x_2} & \frac{\partial f_3}{\partial x_3} & \cdots & \frac{\partial f_3}{\partial x_n} \\[2mm] \vdots & \vdots & \vdots & \vdots & \vdots \\[2mm] \frac{\partial f_n}{\partial x_1} & \frac{\partial f_n}{\partial x_2} & \frac{\partial f_n}{\partial x_3} & \cdots & \frac{\partial f_n}{\partial x_n} \end{bmatrix}$$

Typically the inverse of the Jacobian $\left[J\left(x^k \right) \right]$ is not found directly, but rather through LU factorization by posing the Newton-Raphson method as

$$\left[J\left(x^k \right) \right] \left(x^{k+1} - x^k \right) = -F\left(x^k \right) \tag{3.19}$$

which is now in the form $Ax = b$ where the Jacobian is the matrix A, the function $-F\left(x^k \right)$ is the vector b, and the unknown x is the difference vector $\left(x^{k+1} - x^k \right)$. Convergence is typically evaluated by considering the norm of the function

$$\left\| F\left(x^k \right) \right\| < \varepsilon \tag{3.20}$$

Note that the Jacobian is a function of x^k and is therefore updated every iteration along with $F\left(x^k \right)$.

Example 3.5
Find the solution to

$$0 = x_1^2 + x_2^2 - 5x_1 + 1 = f_1(x_1, x_2) \qquad (3.21)$$
$$0 = x_1^2 - x_2^2 - 3x_2 - 3 = f_2(x_1, x_2) \qquad (3.22)$$

with an initial guess of

$$x^{(0)} = \begin{bmatrix} 3 \\ 3 \end{bmatrix}$$

Solution 3.5 The Jacobian of this system of equations is:

$$J(x_1, x_2) = \begin{bmatrix} \frac{\partial f_1}{\partial x_1} & \frac{\partial f_1}{\partial x_2} \\ \frac{\partial f_2}{\partial x_1} & \frac{\partial f_2}{\partial x_2} \end{bmatrix} = \begin{bmatrix} 2x_1 - 5 & 2x_2 \\ 2x_1 & -2x_2 - 3 \end{bmatrix}$$

Iteration 1
The Jacobian and the functions f_1 and f_2 are evaluated at the initial condition

$$\begin{bmatrix} 1 & 6 \\ 6 & -9 \end{bmatrix} \begin{bmatrix} x_1^{(1)} - 3 \\ x_2^{(1)} - 3 \end{bmatrix} = \begin{bmatrix} -4 \\ 12 \end{bmatrix} \qquad (3.23)$$

Solving this linear system yields

$$\begin{bmatrix} x_1^{(1)} - 3 \\ x_2^{(1)} - 3 \end{bmatrix} = \begin{bmatrix} 0.8 \\ -0.8 \end{bmatrix} \qquad (3.24)$$

Thus

$$x_1^{(1)} = 0.8 + x_1^{(0)} = 0.8 + 3 = 3.8 \qquad (3.25)$$
$$x_2^{(1)} = -0.8 + x_2^{(0)} = -0.8 + 3 = 2.2 \qquad (3.26)$$

The error at iteration 1 is

$$\left\| \begin{bmatrix} f_1\left(x_1^{(0)}, x_2^{(0)}\right) \\ f_2\left(x_1^{(0)}, x_2^{(0)}\right) \end{bmatrix} \right\|_\infty = 12$$

Iteration 2
The Jacobian and the functions f_1 and f_2 are evaluated at $x^{(1)}$

$$\begin{bmatrix} 2.6 & 4.4 \\ 7.6 & -7.4 \end{bmatrix} \begin{bmatrix} x_1^{(2)} - 3.8 \\ x_2^{(2)} - 2.2 \end{bmatrix} = \begin{bmatrix} -1.28 \\ 0.00 \end{bmatrix} \qquad (3.27)$$

Solving this linear system yields

$$\begin{bmatrix} x_1^{(2)} - 3.8 \\ x_2^{(2)} - 2.2 \end{bmatrix} = \begin{bmatrix} -0.1798 \\ -0.1847 \end{bmatrix} \qquad (3.28)$$

Thus

$$x_1^{(2)} = -0.1798 + x_1^{(1)} = -0.1798 + 3.8 = 3.6202 \tag{3.29}$$
$$x_2^{(2)} = -0.1847 + x_2^{(1)} = -0.1847 + 2.2 = 2.0153 \tag{3.30}$$

The error at iteration 2 is

$$\left\| \begin{bmatrix} f_1\left(x_1^{(1)}, x_2^{(1)}\right) \\ f_2\left(x_1^{(1)}, x_2^{(1)}\right) \end{bmatrix} \right\|_\infty = 1.28$$

Iteration 3

The Jacobian and the functions f_1 and f_2 are evaluated at $x^{(2)}$

$$\begin{bmatrix} 2.2404 & 4.0307 \\ 7.2404 & -7.0307 \end{bmatrix} \begin{bmatrix} x_1^{(3)} - 3.6202 \\ x_2^{(3)} - 2.0153 \end{bmatrix} = \begin{bmatrix} -0.0664 \\ 0.0018 \end{bmatrix} \tag{3.31}$$

Solving this linear system yields

$$\begin{bmatrix} x_1^{(3)} - 3.6202 \\ x_2^{(3)} - 2.0153 \end{bmatrix} = \begin{bmatrix} -0.0102 \\ -0.0108 \end{bmatrix} \tag{3.32}$$

Thus

$$x_1^{(3)} = -0.0102 + x_1^{(2)} = -0.0102 + 3.6202 = 3.6100 \tag{3.33}$$
$$x_2^{(3)} = -0.0108 + x_2^{(2)} = -0.0108 + 2.0153 = 2.0045 \tag{3.34}$$

The error at iteration 3 is

$$\left\| \begin{bmatrix} f_1\left(x_1^{(2)}, x_2^{(2)}\right) \\ f_2\left(x_1^{(2)}, x_2^{(2)}\right) \end{bmatrix} \right\|_\infty = 0.0664$$

At iteration 4, the functions f_1 and f_2 are evaluated at $x^{(3)}$ and yield the following:

$$\begin{bmatrix} f_1\left(x_1^{(3)}, x_2^{(3)}\right) \\ f_2\left(x_1^{(3)}, x_2^{(3)}\right) \end{bmatrix} = \begin{bmatrix} -0.221 \times 10^{-3} \\ 0.012 \times 10^{-3} \end{bmatrix}$$

Since the norm of this matrix is very small, it can be concluded that the iterates have converged and

$$\begin{bmatrix} x_1^{(3)} \\ x_2^{(3)} \end{bmatrix} = \begin{bmatrix} 3.6100 \\ 2.0045 \end{bmatrix}$$

are within an order of error of 10^{-3} of the actual solution. ∎

In the solution to Example 3.5, the error at each iteration is

iteration	error
0	12.0000
1	1.2800
2	0.0664
3	0.0002

Note that once the solution is sufficiently close to the actual solution, the error at each iteration decreases rapidly. If the iterations were carried far enough the error at each iteration would become roughly the square of the previous iteration error. This convergence behavior is indicative of the quadratic convergence of the Newton-Raphson method.

3.2.3   Modifications to the Newton-Raphson Method

Although the full Newton-Raphson method exhibits quadratic convergence and a minimum number of iterations, each iteration may require significant computation. For example, the computation of a full Jacobian matrix requires n^2 calculations, and each iteration requires on the order of n^3 operations for the LU factorization if the Jacobian is a full matrix. Therefore most modifications to the Newton-Raphson method propose to reduce either the calculation or the LU factorization of the Jacobian matrix.

Consider once again the iterative statement of the Newton-Raphson method:

$$I: \quad x^{k+1} = x^k - \left[J\left(x^k\right)\right]^{-1} f\left(x^k\right)$$

This iterative statement can be written in a more general form as:

$$I: \quad x^{k+1} = x^k - \left[M\left(x^k\right)\right]^{-1} f\left(x^k\right) \tag{3.35}$$

where M is an $n \times n$ matrix that may or may not be a function of x^k. Note that even if $M \neq J$, this iteration will still converge to a correct solution for x if the function $f(x)$ is driven to zero. So one approach to simplifying the Newton-Raphson method is to find a suitable substitute matrix M that is easier to compute than the system Jacobian. One common simplification is to substitute each of the partial derivative entries $\frac{\partial f_i}{\partial x_j}$ by a difference approximation. For example, a simple approximation might be

$$\frac{\partial f_i}{\partial x_j} \approx \frac{1}{h_{ij}} \left[f_i\left(x + h_{ij}e^j\right) - f_i\left(x\right)\right] \tag{3.36}$$

where e^j is the jth unit vector:

$$e^j = \begin{bmatrix} 0 \\ 0 \\ \vdots \\ 0 \\ 1 \\ 0 \\ \vdots \\ 0 \end{bmatrix}$$

where the 1 occurs in the jth row of the unit vector and all other entries are zero. The scalar h_{ij} can be chosen in numerous ways, but one common choice is to let $h_{ij}^k = x_j^k - x_j^{k-1}$. This choice for h_{ij} leads to a rate of convergence of 1.62, which lies between quadratic and linear convergence rates.

Another common modification to the Newton-Raphson method is to set M equal to the Jacobian matrix at occasional intervals. For example, the matrix M can be re-evaluated whenever the convergence slows down, or at more regular intervals, such as every other or other third iteration. This modification is known as the *dishonest Newton's* method. An extreme extension of this method is to set M equal to the initial Jacobian matrix and then to hold it constant throughout the remainder of the iteration. This is commonly called the *very dishonest Newton's* method. In addition to the reduction in computation associated with the calculation of the matrix, this method also has the advantage that the M matrix need only be factored into the LU matrices once since it is a constant. This can save considerable computation time in the LU factorization process. Similarly, the matrices of the dishonest Newton's method need only be factored when the M matrix is re-evaluated.

3.3 Power System Applications

The solution and analysis procedures outlined in this chapter form the basis of a set of powerful tools that can be used for a myriad of power system applications. One of the most outstanding features of power systems is that they are modeled as an extremely large set of nonlinear equations. The North American transmission grid is one of the largest nonlinear engineering systems. Most types of power system analysis require the solution in one form or another of this system of nonlinear equations. The applications described below are a handful of the more common applications, but are certainly not a complete coverage of all possible nonlinear problems that arise in power system analysis.

3.3.1 Power Flow

Many power system problems give rise to systems of nonlinear equations that must be solved. Probably the most common nonlinear power system problem is the *power flow* or *load flow* problem. The underlying principle of a power flow problem is that given the system loads, generation, and network configuration, the system bus voltages and line flows can be found by solving the nonlinear power flow equations. This is typically accomplished by applying Kirchoff's law at each power system bus throughout the system. In this context, Kirchoff's law can be interpreted as *the sum of the powers entering a bus must be zero,* or that the power at each bus must be conserved. Since power is comprised of two components, active power and reactive power, each bus gives rise to two equations – one for active power and one for reactive power. These equations are known as the *power flow equations:*

$$0 = \Delta P_i = P_i^{inj} - V_i \sum_{j=1}^{N_{bus}} V_j Y_{ij} \cos\left(\theta_i - \theta_j - \phi_{ij}\right) \tag{3.37}$$

$$0 = \Delta Q_i = Q_i^{inj} - V_i \sum_{j=1}^{N_{bus}} V_j Y_{ij} \sin\left(\theta_i - \theta_j - \phi_{ij}\right) \tag{3.38}$$

$$i = 1, \ldots, N_{bus}$$

where P_i^{inj}, Q_i^{inj} are the active and reactive power injected at the bus i, respectively. Loads are modeled by negative power injection. The values V_i and V_j are the voltage magnitudes at bus i and bus j, respectively. The values θ_i and θ_j are the corresponding phase angles. The value $Y_{ij} \angle \phi_{ij}$ is the $(ij)^{th}$ element of the network admittance matrix Y_{bus}. The constant N_{bus} is the number of buses in the system. The updates ΔP_i^k and ΔQ_i^k of equations (3.37) and (3.38) are called the *mismatch* equations because they give a measure of the power difference, or mismatch, between the calculated power values, as functions of voltage and phase angle, and the actual injected powers. As the Newton-Raphson iteration continues, this mismatch is driven to zero until the power leaving a bus, calculated from the voltages and phase angles, equals the injected power. At this point the converged values of voltages and phase angles are used to calculate line flows, swing bus powers, and the injected reactive powers at the generator buses.

The formulation in equations (3.37) and (3.38) is called the *polar* formulation of the power flow equations. If $Y_{ij} \angle \phi_{ij}$ is instead given by the complex sum $g_{ij} + jb_{ij}$, then the power flow equations may be written in *rectangular form* as

$$0 = P_i^{inj} - V_i \sum_{j=1}^{N_{bus}} V_j \left(g_{ij} \cos\left(\theta_i - \theta_j\right) + b_{ij} \sin\left(\theta_i - \theta_j\right)\right) \tag{3.39}$$

$$0 = Q_i^{inj} - V_i \sum_{j=1}^{N_{bus}} V_j \left(g_{ij} \sin (\theta_i - \theta_j) - b_{ij} \cos (\theta_i - \theta_j) \right) \qquad (3.40)$$

$$i, \dots, N_{bus}$$

In either case, the power flow equations are a system of nonlinear equations. They are nonlinear in both the voltage and phase angle. There are, at most, $2N_{bus}$ equations to solve. This number is then further reduced by removing one power flow equation for each known voltage (at voltage controlled buses) and the swing bus angle. This reduction is necessary since the number of equations must equal the number of unknowns in a fully determined system. Once the nonlinear power flow equations have been determined, the Newton-Raphson method may be directly applied.

The most common approach to solving the power flow equations by the Newton-Raphson method is to arrange the equations by phase angle followed by the voltage magnitudes as

$$\begin{bmatrix} J_1 & J_2 \\ J_3 & J_4 \end{bmatrix} \begin{bmatrix} \Delta \delta_1 \\ \Delta \delta_2 \\ \Delta \delta_3 \\ \vdots \\ \Delta \delta_{N_{bus}} \\ \Delta V_1 \\ \Delta V_2 \\ \Delta V_3 \\ \vdots \\ \Delta V_{N_{bus}} \end{bmatrix} = - \begin{bmatrix} \Delta P_1 \\ \Delta P_2 \\ \Delta P_3 \\ \vdots \\ \Delta P_{N_{bus}} \\ \Delta Q_1 \\ \Delta Q_2 \\ \Delta Q_3 \\ \vdots \\ \Delta Q_{N_{bus}} \end{bmatrix} \qquad (3.41)$$

where

$$\Delta \delta_i = \delta_i^{k+1} - \delta_i^k$$
$$\Delta V_i = V_i^{k+1} - V_i^k$$

These equations are then solved using LU factorization with forward/backward substitution. The Jacobian is typically divided into four submatrices, where

$$\begin{bmatrix} J_1 & J_2 \\ J_3 & J_4 \end{bmatrix} = \begin{bmatrix} \frac{\partial \Delta P}{\partial \delta} & \frac{\partial \Delta P}{\partial V} \\ \frac{\partial \Delta Q}{\partial \delta} & \frac{\partial \Delta Q}{\partial V} \end{bmatrix} \qquad (3.42)$$

Each submatrix represents the partial derivatives of each of the mismatch equations with respect to each of the unknowns. These partial derivatives yield eight types – two for each mismatch equation, where one is for the diagonal element and the other is for off-diagonal elements. The derivatives are summarized as

$$\frac{\partial \Delta P_i}{\partial \delta_i} = V_i \sum_{j=1}^{N_{bus}} V_j Y_{ij} \sin (\delta_i - \delta_j - \phi_{ij}) + V_i^2 Y_{ii} \sin \phi_{ii} \qquad (3.43)$$

$$\frac{\partial \Delta P_i}{\partial \delta_j} = -V_i V_j Y_{ij} \sin(\delta_i - \delta_j - \phi_{ij}) \tag{3.44}$$

$$\frac{\partial \Delta P_i}{\partial V_i} = -\sum_{i=1}^{N_{bus}} V_j Y_{ij} \cos(\delta_i - \delta_j - \phi_{ij}) - V_i Y_{ii} \cos \phi_{ii} \tag{3.45}$$

$$\frac{\partial \Delta P_i}{\partial V_j} = -V_i Y_{ij} \cos(\delta_i - \delta_j - \phi_{ij}) \tag{3.46}$$

$$\frac{\partial \Delta Q_i}{\partial \delta_i} = -V_i \sum_{j=1}^{N_{bus}} V_j Y_{ij} \cos(\delta_i - \delta_j - \phi_{ij}) + V_i^2 Y_{ii} \cos \phi_{ii} \tag{3.47}$$

$$\frac{\partial \Delta Q_i}{\partial \delta_j} = V_i V_j Y_{ij} \cos(\delta_i - \delta_j - \phi_{ij}) \tag{3.48}$$

$$\frac{\partial \Delta Q_i}{\partial V_i} = -\sum_{j=1}^{N_{bus}} V_j Y_{ij} \sin(\delta_i - \delta_j - \phi_{ij}) + V_i Y_{ii} \sin \phi_{ii} \tag{3.49}$$

$$\frac{\partial \Delta Q_i}{\partial V_j} = -V_i Y_{ij} \sin(\delta_i - \delta_j - \phi_{ij}) \tag{3.50}$$

A common modification to the power flow solution is to replace the unknown update ΔV_i by the normalized value $\frac{\Delta V_i}{V_i}$. This formulation yields a more symmetric Jacobian as the Jacobian submatrices J_2 and J_4 are now multiplied by V_i to compensate for the scaling of ΔV_i by V_i. All partial derivatives of each submatrix then become quadratic in voltage magnitude.

The Newton-Raphson method for the solution of the power flow equations is relatively straightforward to program since both the function evaluations and the partial derivatives use the same expressions. Thus it takes little extra computational effort to compute the Jacobian once the mismatch equations have been calculated.

Example 3.6

Find the voltage magnitudes, phase angles, and line flows for the small power system shown in Figure 3.7 with the following system parameters in per unit:

bus	type	V	P_{gen}	Q_{gen}	P_{load}	Q_{load}
1	swing	1.02	–	–	0.0	0.0
2	PV	1.00	0.5	–	0.0	0.0
3	PQ	–	0.0	0.0	1.2	0.5

i	j	R_{ij}	X_{ij}	B_{ij}
1	2	0.02	0.3	0.15
1	3	0.01	0.1	0.1
2	3	0.01	0.1	0.1

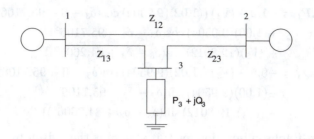

FIGURE 3.7
Example power system

Solution 3.6 The first step in any power flow solution is to calculate the admittance matrix Y_{bus} for the power system. A simple procedure for calculating the elements of the admittance matrix is

Y_{ij}	negative of the admittance between buses i and j
Y_{ii}	sum of all admittances connected to bus i

Calculating the admittance matrix for this system yields:

$$Y_{bus} = \begin{bmatrix} 13.1505\angle - 84.7148° & 3.3260\angle 93.8141° & 9.9504\angle 95.7106° \\ 3.3260\angle 95.7106° & 13.1505\angle - 84.7148° & 9.9504\angle 95.7106° \\ 9.9504\angle 95.7106° & 9.9504\angle 95.7106° & 19.8012\angle - 84.2606° \end{bmatrix}$$
$$(3.51)$$

By inspection, this system has three unknowns: δ_2, δ_3, and V_3; thus, three power flow equations are required. These power flow equations are

$$0 = \Delta P_2 = 0.5 - V_2 \sum_{j=1}^{3} V_j Y_{ij} \cos (\delta_2 - \delta_j - \theta_{ij}) \qquad (3.52)$$

$$0 = \Delta P_3 = -1.2 - V_3 \sum_{j=1}^{3} V_j Y_{ij} \cos (\delta_3 - \delta_j - \theta_{ij}) \qquad (3.53)$$

$$0 = \Delta Q_3 = -0.5 - V_3 \sum_{j=1}^{3} V_j Y_{ij} \sin (\delta_3 - \delta_j - \theta_{ij}) \qquad (3.54)$$

Substituting in the known quantities for $V_1 = 1.02, V_2 = 1.00$, and $\delta_1 = 0$ and the admittance matrix quantities yields:

$$\Delta P_2 = 0.5 - (1.00) \left((1.02)(3.3260) \cos(\delta_2 - 0 - 93.8141°) \right.$$
$$+ (1.00)(13.1505) \cos(\delta_2 - \delta_2 + 84.7148°)$$
$$\left. + (V_3)(9.9504) \cos(\delta_2 - \delta_3 - 95.7106°) \right) \qquad (3.55)$$

$$\Delta P_3 = -1.2 - (V_3) \left((1.02)(9.9504) \cos(\delta_3 - 0 - 95.7106°) \right.$$
$$+ (1.00)(9.9504) \cos(\delta_3 - \delta_2 - 95.7106°)$$
$$+ (V_3)(19.8012) \cos(\delta_3 - \delta_3 + 84.2606°)) \qquad (3.56)$$
$$\Delta Q_3 = -0.5 - (V_3) \left((1.02)(9.9504) \sin(\delta_3 - 0 - 95.7106°) \right.$$
$$+ (1.00)(9.9504) \sin(\delta_3 - \delta_2 - 95.7106°)$$
$$+ ((V_3)(19.8012) \sin(\delta_3 - \delta_3 + 84.2606°)) \qquad (3.57)$$

The Newton-Raphson iteration for this system is then given by

$$
\begin{bmatrix}
\frac{\partial \Delta P_2}{\partial \delta_2} & \frac{\partial \Delta P_2}{\partial \delta_3} & \frac{\partial \Delta P_2}{\partial V_3} \\
\frac{\partial \Delta P_3}{\partial \delta_2} & \frac{\partial \Delta P_3}{\partial \delta_3} & \frac{\partial \Delta P_3}{\partial V_3} \\
\frac{\partial \Delta Q_3}{\partial \delta_2} & \frac{\partial \Delta Q_3}{\partial \delta_3} & \frac{\partial \Delta Q_3}{\partial V_3}
\end{bmatrix}
\begin{bmatrix}
\Delta \delta_2 \\
\Delta \delta_3 \\
\Delta V_3
\end{bmatrix}
= -
\begin{bmatrix}
\Delta P_2 \\
\Delta P_3 \\
\Delta Q_3
\end{bmatrix}
\qquad (3.58)
$$

where

$$\frac{\partial \Delta P_2}{\partial \delta_2} = 3.3925 \sin(\delta_2 - 93.8141°)$$
$$+ 9.9504 V_3 \sin(\delta_2 - \delta_3 - 95.7106°)$$
$$\frac{\partial \Delta P_2}{\partial \delta_3} = -9.9504 V_3 \sin(\delta_2 - \delta_3 - 95.7106°)$$
$$\frac{\partial \Delta P_2}{\partial V_3} = -9.9504 \cos(\delta_2 - \delta_3 - 95.7106°)$$
$$\frac{\partial \Delta P_3}{\partial \delta_2} = -9.9504 V_3 \sin(\delta_3 - \delta_2 - 95.7106°)$$
$$\frac{\partial \Delta P_3}{\partial \delta_3} = 10.1494 V_3 \sin(\delta_3 - 95.7106°)$$
$$+ 9.9504 V_3 \sin(\delta_3 - \delta_2 - 95.7106°)$$
$$\frac{\partial \Delta P_3}{\partial V_3} = -10.1494 \cos(\delta_3 - 95.7106°)$$
$$-9.9504 \cos(\delta_3 - \delta_2 - 95.7106°)$$
$$-39.6024 V_3 \cos(84.2606°)$$
$$\frac{\partial \Delta Q_3}{\partial \delta_2} = 9.9504 V_3 \cos(\delta_3 - \delta_2 - 95.7106°)$$
$$\frac{\partial \Delta Q_3}{\partial \delta_3} = -10.1494 V_3 \cos(\delta_3 - 95.7106°)$$
$$-9.9504 V_3 \cos(\delta_3 - \delta_2 - 95.7106°)$$
$$\frac{\partial \Delta Q_3}{\partial V_3} = -10.1494 \sin(\delta_3 - 95.7106°)$$
$$-9.9504 \sin(\delta_3 - \delta_2 - 95.7106°)$$
$$-39.6024 V_3 \sin(84.2606°)$$

Recall that one of the underlying assumptions of the Newton-Raphson iteration is that the higher order terms of the Taylor series expansion are negligible only if the initial guess is sufficiently close to the actual solution to the nonlinear equations. Under most operating conditions, the voltages throughout the power system are within ±10% of the nominal voltage and therefore fall in the range $0.9 \leq V_i \leq 1.1$ per unit. Similarly, under most operating conditions the phase angle differences between adjacent buses are typically small. Thus if the swing bus angle is taken to be zero, then all phase angles throughout the system will also be close to zero. Therefore in initializing a power flow, it is common to choose a "flat start" initial condition. That is, all voltage magnitudes are set to 1.0 per unit and all angles are set to zero.

Iteration 1

Evaluating the Jacobian and the mismatch equations at the flat start initial conditions yields:

$$[J^0] = \begin{bmatrix} -13.2859 & 9.9010 & 0.9901 \\ 9.9010 & -20.0000 & -1.9604 \\ -0.9901 & 2.0000 & -19.4040 \end{bmatrix}$$

$$\begin{bmatrix} \Delta P_2^0 \\ \Delta P_3^0 \\ \Delta Q_3^0 \end{bmatrix} = \begin{bmatrix} 0.5044 \\ -1.1802 \\ -0.2020 \end{bmatrix}$$

Solving

$$[J^0] \begin{bmatrix} \Delta \delta_2^1 \\ \Delta \delta_3^1 \\ \Delta V_3^1 \end{bmatrix} = - \begin{bmatrix} \Delta P_2^0 \\ \Delta P_3^0 \\ \Delta Q_3^0 \end{bmatrix}$$

by LU factorization yields:

$$\begin{bmatrix} \Delta \delta_2^1 \\ \Delta \delta_3^1 \\ \Delta V_3^1 \end{bmatrix} = \begin{bmatrix} -0.0096 \\ -0.0621 \\ -0.0163 \end{bmatrix}$$

Therefore

$$\delta_2^1 = \delta_2^0 + \Delta \delta_2^1 = 0 - 0.0096 = -0.0096$$
$$\delta_3^1 = \delta_3^0 + \Delta \delta_3^1 = 0 - 0.0621 = -0.0621$$
$$V_3^1 = V_3^0 + \Delta V_3^1 = 1 - 0.0163 = 0.9837$$

Note that the angles are given in *radians* and not degrees. The error at the first iteration is the largest absolute value of the mismatch equations, which is

$$\varepsilon^1 = 1.1802$$

One quick check of this process is to note that the voltage update V_3^1 is slightly less than 1.0 per unit, which would be expected given the system configuration. Note also that the diagonals of the Jacobian are all equal or greater in magnitude than the off-diagonal elements. This is because the diagonals are summations of terms, whereas the off-diagonal elements are single terms.

Iteration 2

Evaluating the Jacobian and the mismatch equations at the updated values δ_2^1, δ_3^1, and V_3^1 yields:

$$[J^1] = \begin{bmatrix} -13.1597 & 9.7771 & 0.4684 \\ 9.6747 & -19.5280 & -0.7515 \\ -1.4845 & 3.0929 & -18.9086 \end{bmatrix}$$

$$\begin{bmatrix} \Delta P_2^1 \\ \Delta P_3^1 \\ \Delta Q_3^1 \end{bmatrix} = \begin{bmatrix} 0.0074 \\ -0.0232 \\ -0.0359 \end{bmatrix}$$

Solving for the update yields

$$\begin{bmatrix} \Delta \delta_2^2 \\ \Delta \delta_3^2 \\ \Delta V_3^2 \end{bmatrix} = \begin{bmatrix} -0.0005 \\ -0.0014 \\ -0.0021 \end{bmatrix}$$

and

$$\begin{bmatrix} \delta_2^2 \\ \delta_3^2 \\ V_3^2 \end{bmatrix} = \begin{bmatrix} -0.0101 \\ -0.0635 \\ 0.9816 \end{bmatrix}$$

where

$$\varepsilon^2 = 0.0359$$

Iteration 3

Evaluating the Jacobian and the mismatch equations at the updated values δ_2^2, δ_3^2, and V_3^2 yields:

$$[J^2] = \begin{bmatrix} -13.1392 & 9.7567 & 0.4600 \\ 9.6530 & -19.4831 & -0.7213 \\ -1.4894 & 3.1079 & -18.8300 \end{bmatrix}$$

$$\begin{bmatrix} \Delta P_2^0 \\ \Delta P_3^0 \\ \Delta Q_3^0 \end{bmatrix} = \begin{bmatrix} 0.1717 \\ -0.5639 \\ -0.9084 \end{bmatrix} \times 10^{-4}$$

Solving for the update yields

$$\begin{bmatrix} \Delta \delta_2^2 \\ \Delta \delta_3^2 \\ \Delta V_3^2 \end{bmatrix} = \begin{bmatrix} -0.1396 \\ -0.3390 \\ -0.5273 \end{bmatrix} \times 10^{-5}$$

and

$$\begin{bmatrix} \delta_2^3 \\ \delta_3^3 \\ V_3^3 \end{bmatrix} = \begin{bmatrix} -0.0101 \\ -0.0635 \\ 0.9816 \end{bmatrix}$$

where

$$\varepsilon^3 = 0.9084 \times 10^{-4}$$

At this point, the iterations have converged since the mismatch is sufficiently small and the values are no longer changing significantly.

The last task in power flow is to calculate the generated reactive powers, the swing bus active power output and the line flows. The generated powers can be calculated directly from the power flow equations:

$$P_i^{inj} = V_i \sum_{j=1}^{N_{bus}} V_j Y_{ij} \cos{(\theta_i - \theta_j - \phi_{ij})}$$

$$Q_i^{inj} = V_i \sum_{j=1}^{N_{bus}} V_j Y_{ij} \sin{(\theta_i - \theta_j - \phi_{ij})}$$

Therefore

$$P_{gen,1} = P_1^{inj} = 0.7087$$
$$Q_{gen,1} = Q_1^{inj} = 0.2806$$
$$Q_{gen,2} = Q_2^{inj} = -0.0446$$

The active power losses in the system are the difference between the sum of the generation and the sum of the loads, in this case:

$$P_{loss} = \sum P_{gen} - \sum P_{load} = 0.7087 + 0.5 - 1.2 = 0.0087 \ pu \qquad (3.59)$$

The line losses for line $i-j$ are calculated at both the sending and receiving ends of the line. Therefore the power sent from bus i to bus j is

$$S_{ij} = V_i \angle \delta_i I_{ij}^* \qquad (3.60)$$

and the power received at bus j from bus i is

$$S_{ji} = V_j \angle \delta_j I_{ji}^* \qquad (3.61)$$

Thus

$$P_{ij} = V_i V_j Y_{ij} \cos{(\delta_i - \delta_j - \phi_{ij})} - V_i^2 Y_{ij} \cos{(\phi_{ij})} \qquad (3.62)$$
$$Q_{ij} = V_i V_j Y_{ij} \sin{(\delta_i - \delta_j - \phi_{ij})} + V_i^2 Y_{ij} \sin{(\phi_{ij})} \qquad (3.63)$$

Similarly, the powers P_{ji} and Q_{ji} can be calculated. The active power loss on any given line is the difference between the active power sent from bus i and the active power received at bus j. Calculating the reactive power losses is more complex since the reactive power generated by the line-charging (shunt capacitances) must also be included. ∎

3.3.2　Regulating Transformers

One of the most common controllers found in the power system network is the *regulating transformer*. This is a transformer that is able to change the winding ratios (tap settings) in response to changes in load-side voltage. If the voltage on the secondary side (or load side) is lower than a desired voltage (such as during heavy loading), the tap will change so as to increase the secondary voltage while maintaining the primary side voltage. A regulating transformer is also frequently referred to as an *under-load-tap-changing* or ULTC transformer. The tap setting t may be real or complex, and in per unit, the tap ratio is defined as $1 : t$ where t is typically within 10% of 1.0. A phase-shifting transformer is achieved by allowing the tap t to be complex with both magnitude and angle.

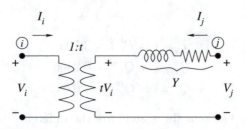

FIGURE 3.8
A regulating transformer

The effect of the regulating transformer is incorporated into the power flow algorithm through the admittance matrix. To incorporate a regulating transformer into the admittance matrix, consider the regulating transformer as a two-port network relating the input currents I_i and I_j to the input voltages V_i and V_j as shown in Figure 3.8. The receiving end current is given by

$$I_j = (V_j - tV_i)\, Y \tag{3.64}$$

Note that the currents can be found from the power transfer equation:

$$S_i = V_i I_i^* = -tV_i I_j^* \tag{3.65}$$

Therefore

$$I_i = -t^* I_j \tag{3.66}$$
$$= -t^* (V_j - tV_i)\, Y \tag{3.67}$$
$$= tt^* Y V_i - t^* Y V_j \tag{3.68}$$
$$= |t|^2\, Y V_i - t^* Y V_j \tag{3.69}$$

Therefore the off-diagonal entries in the admittance matrix become:

$$Y(i,j) = -t^*Y$$
$$Y(j,i) = -tY$$

and $|t|^2 Y$ is added to $Y(i,i)$ and Y is added to $Y(j,j)$.

Since regulating transformers are used as voltage control devices, a common computational exercise is to find the tap setting t that will hold the secondary bus voltage magnitude V_j at a specified voltage $\hat{V}$. This may be interpreted as adding one additional variable to the system (t) and one additional constraint $\left(V_j = \hat{V}\right)$. Since the additional constraint is counterbalanced by the additional degree of freedom, the dimension of the problem remains the same. There are two primary approaches for finding the tap setting t that results in $V_j = \hat{V}$. One approach is an iterative approach while the second approach calculates t directly from the power flow equations.

The iterative approach may be summarized as:

1. Set $t = t_0$

2. Run a power flow to calculate V_j

3. Is $V_j > \hat{V}$? If yes, then $t = t - \Delta t$, and go to step 2.

4. Is $V_j < \hat{V}$? If yes, then $t = t + \Delta t$, and go to step 2.

5. Done

This approach is conceptually simple and requires no changes to the power flow algorithm. However, it may require numerous runs of a power flow program if t_0 is far from the required tap setting.

The direct approach applies the Newton-Raphson method directly to the updated power flow equations as functions of the tap setting t.

1. Set $V_j = \hat{V}$ and let t be an unknown state

2. Modify the Newton-Raphson Jacobian such that the row of partial derivatives with respect to V_j are replaced by the row of partial derivatives with respect to t

3. Modify the state vector x such that

$$x = \begin{bmatrix} \delta_2 \\ \delta_3 \\ \vdots \\ \delta_n \\ V_2 \\ V_3 \\ \vdots \\ V_{j-1} \\ t \\ V_{j+1} \\ \vdots \\ V_n \end{bmatrix}$$

Note that the state V_j is replaced by t.

4. Perform the Newton-Raphson

In this case, the set of power flow equations is solved only once, but since the system Jacobian is modified, a standard power flow program cannot be used.

Since the tap cannot move continuously along the transformer windings, but must move vertically from one winding to the adjacent winding, the real tap setting is not a continuous state. Therefore, in both cases, the calculated tap setting must be rounded to the nearest possible physical tap setting.

Example 3.7
For the system shown in Figure 3.7, place a transformer with reactance X and real tap t between bus 3 and the load (introduce a new bus 4). Find the new admittance matrix and the corresponding Jacobian entries.

Solution 3.7 Let the admittance matrix of the subsystem containing buses 1-3 be given by:

$$Y_{bus} = \begin{bmatrix} Y_{11}\angle\theta_{11} & Y_{12}\angle\theta_{12} & Y_{13}\angle\theta_{13} \\ Y_{21}\angle\theta_{21} & Y_{22}\angle\theta_{22} & Y_{23}\angle\theta_{23} \\ Y_{31}\angle\theta_{31} & Y_{32}\angle\theta_{32} & Y_{33}\angle\theta_{33} \end{bmatrix} \tag{3.70}$$

Adding the transformer between buses 3 and 4 yields the new admittance matrix:

$$Y_{bus} = \begin{bmatrix} Y_{11}\angle\theta_{11} & Y_{12}\angle\theta_{12} & Y_{13}\angle\theta_{13} & 0 \\ Y_{21}\angle\theta_{21} & Y_{22}\angle\theta_{22} & Y_{23}\angle\theta_{23} & 0 \\ Y_{31}\angle\theta_{31} & Y_{32}\angle\theta_{32} & Y_{33}\angle\theta_{33} + \frac{t^2}{jX} & \frac{-t}{jX} \\ 0 & 0 & \frac{-t}{jX} & \frac{1}{jX} \end{bmatrix} \tag{3.71}$$

The power flow equations at bus 3 become:

$$0 = P_3 - V_3 V_1 Y_{31} \cos(\delta_3 - \delta_1 - \theta_{31}) - V_3 V_2 Y_{32} \cos(\delta_3 - \delta_2 - \theta_{32})$$
$$- V_3 V_4 \left(\frac{t}{X}\right) \cos(\delta_3 - \delta_4 - 90°) - V_3^2 Y_{33} \cos(-\theta_{33}) - V_3^2 \left(\frac{t^2}{X}\right) \cos(90°)$$
$$0 = Q_3 - V_3 V_1 Y_{31} \sin(\delta_3 - \delta_1 - \theta_{31}) - V_3 V_2 Y_{32} \sin(\delta_3 - \delta_2 - \theta_{32})$$
$$- V_3 V_4 \left(\frac{t}{X}\right) \sin(\delta_3 - \delta_4 - 90°) - V_3^2 Y_{33} \sin(-\theta_{33}) - V_3^2 \left(\frac{t^2}{X}\right) \sin(90°)$$

Since V_4 is specified, there is no partial derivative $\frac{\partial \Delta P_3}{\partial V_4}$; instead there is a partial derivative with respect to t:

$$\frac{\partial \Delta P_3}{\partial t} = -\frac{V_3 V_4}{X} \cos(\delta_3 - \delta_4 - 90°) \tag{3.72}$$

Similarly, the partial derivative of $\frac{\partial \Delta Q_3}{\partial t}$ becomes

$$\frac{\partial \Delta Q_3}{\partial t} = -\frac{V_3 V_4}{X} \sin(\delta_3 - \delta_4 - 90°) + 2V_3^2 \frac{t}{X} \tag{3.73}$$

The partial derivatives with respect to δ_1, δ_2, V_1, and V_2 do not change, but the partial derivatives with respect to δ_3, δ_4, and V_3 become

$$\frac{\partial \Delta P_3}{\partial \delta_3} = V_3 V_1 Y_{31} \sin(\delta_3 - \delta_1 - \theta_{31}) + V_3 V_2 Y_{32} \sin(\delta_3 - \delta_2 - \theta_{32})$$
$$+ V_3 V_4 \frac{t}{X} \sin(\delta_3 - \delta_4 - 90°)$$

$$\frac{\partial \Delta P_3}{\delta_4} = -V_3 V_4 \frac{t}{X} \sin(\delta_3 - \delta_4 - 90°)$$

$$\frac{\partial \Delta P_3}{\partial V_3} = -V_1 Y_{31} \cos(\delta_3 - \delta_1 - \theta_{31}) - V_2 Y_{32} \cos(\delta_3 - \delta_2 - \theta_{32})$$
$$V_4 \frac{t}{X} \cos(\delta_3 - \delta_4 - 90°) - 2V_3 Y_{33} \cos(-\theta_{33})$$

$$\frac{\partial \Delta Q_3}{\partial \delta_3} = -V_3 V_1 Y_{31} \cos(\delta_3 - \delta_1 - \theta_{31}) - V_3 V_2 Y_{32} \cos(\delta_3 - \delta_2 - \theta_{32})$$
$$- V_3 V_4 \frac{t}{X} \cos(\delta_3 - \delta_4 - 90°)$$

$$\frac{\partial \Delta Q_3}{\partial \delta_4} = V_3 V_4 \frac{t}{X} \cos(\delta_3 - \delta_4 - 90°)$$

$$\frac{\partial \Delta Q_3}{\partial V_3} = -V_1 Y_{31} \sin(\delta_3 - \delta_1 - \theta_{31}) - V_2 Y_{32} \sin(\delta_3 - \delta_2 - \theta_{32})$$
$$- V_4 \frac{t}{X} \sin(\delta_3 - \delta_4 - 90°) - 2V_3 Y_{33} \sin(-\theta_{33}) - 2V_3 \frac{t^2}{X}$$

These partial derivatives are used in developing the Newton-Raphson Jacobian for the iterative power flow method. ∎

3.3.3   Decoupled Power Flow

The power flow is one of the most widely used computational tools in power systems analysis. It can be successfully applied to problems ranging from a single machine system to a power system containing tens of thousands of buses. For very large systems, the full power flow may require significant computational resources to calculate, store, and factorize the Jacobian matrix. As discussed previously, however, it is possible to replace the Jacobian matrix with a matrix M that is easier to calculate and factor and still retain good convergence properties. The power flow equations naturally lend themselves to several alternate matrices for the power flow solution that can be derived from the formulation of the system Jacobian. Recall that the system Jacobian has the form:

$$\begin{bmatrix} J_1 & J_2 \\ J_3 & J_4 \end{bmatrix} = \begin{bmatrix} \frac{\partial \Delta P}{\partial \delta} & \frac{\partial \Delta P}{\partial V} \\ \frac{\partial \Delta Q}{\partial \delta} & \frac{\partial \Delta Q}{\partial V} \end{bmatrix} \tag{3.74}$$

The general form of the P submatrices are:

$$\frac{\partial \Delta P_i}{\partial \delta_j} = -V_i V_j Y_{ij} \sin(\delta_i - \delta_j - \phi_{ij}) \tag{3.75}$$

$$\frac{\partial \Delta P_i}{\partial V_j} = V_i Y_{ij} \cos(\delta_i - \delta_j - \phi_{ij}) \tag{3.76}$$

For most transmission lines, the line resistance contributes only nominally to the overall line impedance; thus, the phase angles ϕ_{ij} of the admittance matrix entries are near $\pm 90°$. Additionally, under normal operating conditions the phase angle difference between adjacent buses is typically small; therefore:

$$\cos(\delta_i - \delta_j - \phi_{ij}) \approx 0 \tag{3.77}$$

leading to

$$\frac{\partial \Delta P_i}{\partial V_j} \approx 0 \tag{3.78}$$

Similar arguments can be made such that

$$\frac{\partial \Delta Q_i}{\partial \delta_j} \approx 0 \tag{3.79}$$

Using the approximations of equations (3.78) and (3.79), a possible substitution for the Jacobian matrix is the matrix

$$M = \begin{bmatrix} \frac{\partial \Delta P}{\partial \delta} & 0 \\ 0 & \frac{\partial \Delta Q}{\partial V} \end{bmatrix} \tag{3.80}$$

Using this matrix M as a replacement for the system Jacobian leads to a set of *decoupled* iterates for the power flow solution:

$$\delta^{k+1} = \delta^k - \left[\frac{\partial \Delta P}{\partial \delta}\right]^{-1} \Delta P \tag{3.81}$$

$$V^{k+1} = V^k - \left[\frac{\partial \Delta Q}{\partial V}\right]^{-1} \Delta Q \tag{3.82}$$

where the ΔP and ΔQ iterations can be carried out independently. The primary advantage of this decoupled power flow is that the LU factorization computation is significantly reduced. The LU factorization of the full Jacobian requires $(2n)^3 = 8n^3$ floating point operations per iteration, whereas the decoupled power flow requires only $2n^3$ floating point operations per iteration.

Example 3.8
Repeat Example 3.6 using the decoupled power flow algorithm.

Solution 3.8 The Jacobian of Example 3.6 evaluated at the initial condition is

$$[J^0] = \begin{bmatrix} -13.2859 & 9.9010 & 0.9901 \\ 9.9010 & -20.0000 & -1.9604 \\ -0.9901 & 2.0000 & -19.4040 \end{bmatrix} \tag{3.83}$$

Note that the off-diagonal submatrices are much smaller in magnitude than the diagonal submatrices. For example,

$$\|[J_2]\| = \left\| \begin{bmatrix} 0.9901 \\ -1.9604 \end{bmatrix} \right\| << \|[J_1]\| = \left\| \begin{bmatrix} -13.2859 & 9.9010 \\ 9.9010 & -20.000 \end{bmatrix} \right\|$$

and

$$\|[J_3]\| = \left\| \begin{bmatrix} -0.9901 & 2.0000 \end{bmatrix} \right\| << \|[J_4]\| = \|[-19.4040]\|$$

Thus, it is reasonable to neglect the off-diagonal matrices J_2 and J_3. Therefore, the first iteration of the decoupled power flow becomes:

$$\begin{bmatrix} \Delta \delta_2^1 \\ \Delta \delta_3^1 \end{bmatrix} = [J_1]^{-1} \begin{bmatrix} \Delta P_2 \\ \Delta P_3 \end{bmatrix} \tag{3.84}$$

$$= \begin{bmatrix} -13.2859 & 9.9010 \\ 9.9010 & -20.000 \end{bmatrix}^{-1} \begin{bmatrix} 0.5044 \\ -1.1802 \end{bmatrix} \tag{3.85}$$

$$[\Delta V_3^1] = [J_4]^{-1} \Delta Q_3 \tag{3.86}$$

$$= -19.4040^{-1} (-0.2020) \tag{3.87}$$

leading to the updates

$$\begin{bmatrix} \delta_2^1 \\ \delta_3^1 \\ V_3^1 \end{bmatrix} = \begin{bmatrix} -0.0095 \\ -0.0637 \\ 0.9896 \end{bmatrix}$$

The iterative process continues similar to the full Newton-Raphson method by continually updating the J_1 and J_4 Jacobian submatrices and the mismatch equations. The iteration converges when both the ΔP mismatch equations and the ΔQ mismatch equations are both less than the convergence tolerance. Note that it is possible for one set of mismatch equations to meet the convergence criteria before the other; thus, the number of "P" iterations required for convergence may differ from the number of "Q" iterations required for convergence. ∎

3.3.4   Fast Decoupled Power Flow

In Example 3.8, each of the decoupled Jacobian submatrices is updated at every iteration. As discussed previously, however, it is often desirable to have constant matrices to minimize the number of function evaluations and LU factorizations. This is often referred to as the *fast decoupled power flow* and can be represented as:

$$[\Delta P^k] = [B'] [\Delta \delta^{k+1}] \tag{3.88}$$

$$\left[\frac{\Delta Q^k}{V} \right] = [B''] [\Delta V^{k+1}] \tag{3.89}$$

where the B' and B'' are constant [35]. To derive these matrices from the power flow Jacobian, consider the decoupled power flow relationships for the Newton-Raphson method:

$$[\Delta P] = - [J_1] [\Delta \delta] \tag{3.90}$$

$$\left[\frac{\Delta Q}{V} \right] = - [J_4] [\Delta V] \tag{3.91}$$

where the Jacobian submatrices in rectangular form are:

$$J_1(i,i) = V_i \sum_{j \neq i} V_j \left(g_{ij} \sin \delta_{ij} - b_{ij} \cos \delta_{ij} \right) \tag{3.92}$$

$$J_1(i,j) = -V_i V_j \left(g_{ij} \sin \delta_{ij} - b_{ij} \cos \delta_{ij} \right) \tag{3.93}$$

$$J_4(i,i) = 2V_i b_{ii} - \sum_{j \neq i} V_j \left(g_{ij} \sin \delta_{ij} - b_{ij} \cos \delta_{ij} \right) \tag{3.94}$$

$$J_4(i,j) = -V_i \left(g_{ij} \sin \delta_{ij} - b_{ij} \cos \delta_{ij} \right) \tag{3.95}$$

where $b_{ij} = |Y_{ij} \sin \phi_{ij}|$ are the imaginary elements of the admittance matrix and $g_{ij} = |Y_{ij} \cos \phi_{ij}|$ are the real elements of the admittance matrix. By noting that $\phi_{ij} \approx 90°$, then $\cos \phi_{ij} \approx 0$ which implies that $g_{ij} \approx 0$. By further approximating all voltage magnitudes as 1.0 per unit, then

$$J_1(i,i) = - \sum_{j \neq i} b_{ij} \tag{3.96}$$

$$J_1(i,j) = b_{ij} \tag{3.97}$$

$$J_4(i,i) = 2b_{ii} - \sum_{j \neq i} b_{ij} \tag{3.98}$$

$$J_4(i,j) = b_{ij} \tag{3.99}$$

Since the J_1 submatrix relates the changes in active power to changes in angle, elements that affect mainly reactive power flow can be omitted from this matrix with negligible impact on the convergence properties. Thus, shunt capacitors (including line-charging) and external reactances as well as the shunts formed due to representation of off-nominal non-phase-shifting transformers (i.e., taps are set to 1.0) are neglected. Hence, the admittance matrix diagonal elements are devoid of these shunts. Additionally, the lumped series resistances of the transmission lines are also omitted. The resulting approximate matrix B' to the submatrix J_1 is given by

$$B'_{ij} = \frac{1}{x_{ij}} \tag{3.100}$$

$$B'_{ii} = -\sum_{j \neq i} B'_{ij} \tag{3.101}$$

Similarly, the J_4 submatrix relates the changes in reactive power to changes in voltage magnitude; therefore elements that primarily affect active power flow are omitted. Thus all phase-shifting transformers are neglected, resulting in

$$B''_{ij} = b_{ij} \tag{3.102}$$

$$B''_{ii} = 2b_i - \sum_{j \neq i} B''_{ij} \tag{3.103}$$

where b_i is the shunt susceptance at bus i (i.e., the sum of susceptances of all the shunt branches connected to bus i).

This method results in a set of constant matrices that can be used to approximate the power flow Jacobian in the Newton-Raphson iteration. This method is often referred to as the XB version of the fast decoupled power flow. Both B' and B'' are real, sparse, and contain only network or admittance matrix elements. In the Newton-Raphson method, these matrices are only factorized once for the LU factorization, and are then stored and held constant throughout the iterative solution process. These matrices were derived based on the application of certain assumptions. If these assumptions do not hold (i.e., the voltage magnitudes deviate substantially from 1.0 per unit; the network has high R/X ratios; or the angle differences between adjacent buses are not small), then convergence problems with the fast decoupled power flow iterations can arise. Work still continues on developing modifications to the XB method to improve convergence [24], [25], [29].

Example 3.9
Repeat Example 3.6 using the fast decoupled power flow algorithm.

Solution 3.9 The line data for the example system are repeated below for convenience:

i	j	R_{ij}	X_{ij}	B_{ij}
1	2	0.02	0.3	0.15
1	3	0.01	0.1	0.1
2	3	0.01	0.1	0.1

and lead to the following admittance matrix:

$$Y_{bus} = \begin{bmatrix} 13.1505\angle -84.7148° & 3.3260\angle 93.8141° & 9.9504\angle 95.7106° \\ 3.3260\angle 95.7106° & 13.1505\angle -84.7148° & 9.9504\angle 95.7106° \\ 9.9504\angle 95.7106° & 9.9504\angle 95.7106° & 19.8012\angle -84.2606° \end{bmatrix}$$

(3.104)

Taking the imaginary part of this matrix yields the following B matrix:

$$B = \begin{bmatrix} -13.0946 & 3.3186 & 9.9010 \\ 3.3186 & -13.0946 & 9.9010 \\ 9.9010 & 9.9010 & -19.7020 \end{bmatrix}$$

(3.105)

From the line data and the associated B matrix, the following B' and B'' matrices result:

$$B' = \begin{bmatrix} -\frac{1}{x_{21}} - \frac{1}{x_{23}} & \frac{1}{x_{23}} \\ \frac{1}{x_{23}} & -\frac{1}{x_{31}} - \frac{1}{x_{32}} \end{bmatrix} = \begin{bmatrix} -13.3333 & 10 \\ 10 & -20 \end{bmatrix}$$

(3.106)

$$B'' = [2b_3 - (B_{31} + B_{32})]$$
$$= [2\,(0.05 + 0.05)) - (9.9010 + 9.9010)] = -19.6020$$

(3.107)

Compare these matrices to the J_1 and J_4 submatrices of Example 3.6 evaluated at the initial condition:

$$J_1 = \begin{bmatrix} -13.2859 & 9.9010 \\ 9.9010 & -20.000 \end{bmatrix}$$

$$J_4 = [-19.4040]$$

The similarity between the matrices is to be expected as a result of the defining assumptions of the fast decoupled power flow method.

Iteration 1

The updates can be found by solving the following linear set of equations

$$\begin{bmatrix} \Delta P_2^0 \\ \Delta P_3^0 \end{bmatrix} = \begin{bmatrix} 0.5044 \\ -1.1802 \end{bmatrix} = -\begin{bmatrix} -13.3333 & 10 \\ 10 & -20 \end{bmatrix} \begin{bmatrix} \Delta \delta_2^1 \\ \Delta \delta_3^1 \end{bmatrix}$$

$$[\Delta Q_3^0] = [-0.2020] = -19.6020 \Delta V_3^1$$

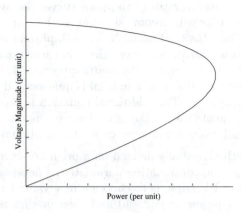

FIGURE 3.9
A PV curve

where $\Delta\delta_2^1 = \delta_2^{(1)} - \delta_2^{(0)}$, $\Delta\delta_3^1 = \delta_3^{(1)} - \delta_3^{(0)}$, and $\Delta V_3^1 = V_3^{(1)} - V_3^{(0)}$ and the initial conditions are a "flat start." Solving for the updates yields

$$\begin{bmatrix} \delta_2^1 \\ \delta_3^1 \\ V_3^1 \end{bmatrix} = \begin{bmatrix} -0.0103 \\ -0.0642 \\ 0.9897 \end{bmatrix}$$

where the phase angles are in radians. This process is continued until convergence in both the "P" and "Q" iterations is achieved. ∎

Note that in both the decoupled power flow cases that the objective of the iterations are the same as for the full Newton-Raphson power flow algorithm. The objective is to drive the mismatch equations ΔP and ΔQ to within some tolerance. Therefore, regardless of the number of iterations required to achieve convergence, the accuracy of the answer is the same as for the full Newton-Raphson method. In other words, the voltages and angles of the decoupled power flow methods will be the same as with the full Newton-Raphson method as long as the iterates converge.

3.3.5 PV Curves and Continuation Power Flow

The power flow is a useful tool for monitoring system voltages as a function of load change. One common application is to plot the voltage at a particular bus as the load is varied from the base case to a loadability limit (often known as the point of maximum loadability). If the load is increased to the loadability limit and then decreased back to the original loading, it is possible to trace the entire power-voltage or "PV" curve. This curve, shown in Figure 3.9, is sometimes called the *nose curve* for its shape.

At the loadability limit, or tip of the nose curve, the system Jacobian of the power flow equations will become singular as the slope of the nose curve becomes infinite. Thus, the traditional Newton-Raphson method of obtaining the load flow solution will break down. In this case, a modification of the Newton-Raphson method known as the *continuation method* is employed. The continuation method introduces an additional equation and unknown into the basic power flow equations. The additional equation is chosen specifically to ensure that the augmented Jacobian is no longer singular at the loadability limit. The additional unknown is often called the continuation parameter.

Continuation methods usually depend on a predictor-corrector scheme and the means to change the continuation parameter as necessary. The basic approach to tracing the PV curve is to choose a new value for the continuation parameter (either in power or voltage) and then predict the power flow solution for this value. This is frequently accomplished using a tangential (or linear) approximation. Using the predicted value as the initial condition for the nonlinear iteration, the augmented power flow equations are then solved (or corrected) to achieve the solution. So the solution is first predicted, and then corrected. This prediction/correction step is shown in Figure 3.10.

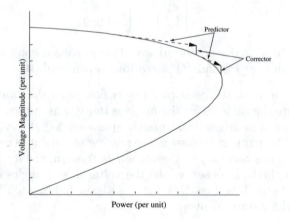

FIGURE 3.10
The predictor/corrector step

Let the set of power flow equations be given as

$$\lambda K - f(\delta, V) = 0 \tag{3.108}$$

or

$$F(\delta, V, \lambda) = 0 \tag{3.109}$$

where K is the loading profile (i.e., the base case relationship between P and Q) and λ is the loading parameter which will vary from unity (at the base case) to the point of maximum loadability. Equation (3.109) may be linearized to yield:

$$\frac{\partial F}{\partial \delta} d\delta + \frac{\partial F}{\partial V} dV + \frac{\partial F}{\partial \lambda} d\lambda = 0 \tag{3.110}$$

Equation (3.110) has one more unknown (λ) than equations, so one more equation is required:

$$e_k \begin{bmatrix} d\delta \\ dV \\ d\lambda \end{bmatrix} = \pm 1 \tag{3.111}$$

where e_k is a row vector of zeros with a single 1 at the position of the unknown that is chosen to be the continuation parameter. The sign of the one on the right hand side is chosen based on whether the continuation parameter is increasing or decreasing. When the continuation parameter is λ (power), the sign is positive indicating that the load is increasing. When voltage is the continuation parameter, the sign is negative, indicating that the voltage magnitude is expected to decrease towards the tip of the nose curve.

The unknowns are predicted such that

$$\begin{bmatrix} \delta \\ V \\ \lambda \end{bmatrix}^{\text{predicted}} = \begin{bmatrix} \delta_0 \\ V_0 \\ \lambda_0 \end{bmatrix} + \sigma \begin{bmatrix} d\delta \\ dV \\ d\lambda \end{bmatrix} \tag{3.112}$$

where

$$\begin{bmatrix} d\delta \\ dV \\ \cdots \\ d\lambda \end{bmatrix} = \begin{bmatrix} & \vdots & \\ J_{LF} & \vdots & K \\ & \vdots & \\ \cdots & \cdots & \cdots \cdots \\ & [e_k] & \end{bmatrix}^{-1} \begin{bmatrix} 0 \\ \vdots \\ 0 \\ 1 \end{bmatrix}$$

and σ is the step-size (or length) for the next prediction. Note that the continuation state $dx_k = 1$, thus

$$x_k^{\text{predicted}} = x_{k0} + \sigma$$

so σ should be chosen to represent a reasonable step-size in terms of what the continuation parameter is (usually voltage or power).

The corrector step involves the solution of the set of equations:

$$F(\delta, V, \lambda) = 0 \tag{3.113}$$

$$x_k - x_k^{\text{predicted}} = 0 \tag{3.114}$$

where x_k is the chosen continuation parameter. Typically the continuation parameter is chosen as the state that exhibits the greatest rate of change.

Example 3.10

Plot the PV curve (P versus V) of the system shown in Figure 3.11 using the continuation power flow method as the load varies from zero to the point of maximum loadability.

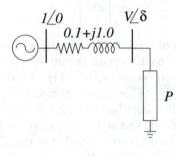

FIGURE 3.11
System for Example 3.10

Solution 3.10 The power flow equations for the system shown in Figure 3.11 are:

$$0 = -P - 0.995V \cos(\delta - 95.7°) - 0.995V^2 \cos(84.3°) \qquad (3.115)$$
$$0 = -0.995V \sin(\delta - 95.7°) - 0.995V^2 \sin(84.3°) \qquad (3.116)$$

During the continuation power flow, the vector of injected active and reactive powers will be replaced by the vector λK. The loading vector λK is

$$\lambda K = \lambda \begin{bmatrix} -1 \\ 0 \end{bmatrix},$$

where λ will vary from zero to the maximum loading value. Typically the vector K will contain the base case values for all injected active and reactive powers in the system. In this case, the entry for the load P is negative indicating that the injected power is negative (i.e., a load).

The loadflow Jacobian for this set of power flow equations is

$$J_{LF} = \begin{bmatrix} 0.995V \sin(\delta - 95.7°) & -0.995 \cos(\delta - 95.7°) - 1.99 \cos(84.3°) V \\ -0.995V \cos(\delta - 95.7°) & -0.995 \sin(\delta - 95.7°) - 1.99 \sin(84.3°) V \end{bmatrix}$$

Iteration 1

Initially, the continuation parameter is chosen to be λ since the load will change more rapidly than the voltage at points far from the tip of the nose curve. At $\lambda = 0$, the circuit is under no-load and the initial voltage magnitude and angle are $1\angle 0°$. With $\sigma = 0.1$ pu, the predictor step yields:

$$\begin{bmatrix} d\delta \\ dV \\ d\lambda \end{bmatrix} = \begin{bmatrix} d\delta_0 \\ dV_0 \\ d\lambda_0 \end{bmatrix} + \sigma \begin{bmatrix} & \vdots & \\ J_{LF} & \vdots & K \\ & \vdots & \\ \cdots & \cdots & \cdots \\ & [e_k] & \end{bmatrix}^{-1} \begin{bmatrix} 0 \\ \vdots \\ 0 \\ 1 \end{bmatrix} \tag{3.117}$$

$$= \begin{bmatrix} 0 \\ 1 \\ 0 \end{bmatrix} + \sigma \begin{bmatrix} -0.9901 & -0.0988 & -1 \\ 0.0988 & -0.9901 & 0 \\ 0 & 0 & 1 \end{bmatrix}^{-1} \begin{bmatrix} 0 \\ 0 \\ 1 \end{bmatrix} \tag{3.118}$$

$$= \begin{bmatrix} -0.1000 \\ 0.9900 \\ 0.1000 \end{bmatrix} \tag{3.119}$$

where δ is in radians. Note that the predicted value for λ is 0.1 pu.

The corrector step solves the system of equations:

$$0 = -\lambda - 0.995V \cos(\delta - 95.7°) - 0.995V^2 \cos(84.3°) \tag{3.120}$$
$$0 = -0.995V \sin(\delta - 95.7°) - 0.995V^2 \sin(84.3°) \tag{3.121}$$

with the load parameter λ set to 0.1 pu. Note that this is a regular loadflow problem and can be solved without program modification.

The first corrector step yields

$$\begin{bmatrix} \delta \\ V \\ \lambda \end{bmatrix} = \begin{bmatrix} -0.1017 \\ 0.9847 \\ 0.1000 \end{bmatrix}$$

Note that this procedure is consistent with the illustration in Figure 3.10. The prediction step is of length σ taken tangentially to the PV at the current point. The corrector step will then occur along a vertical path because the power (λK) is held constant during the correction.

Iteration 2

The second iteration proceeds as the first. The predictor step yields the following guess:

$$\begin{bmatrix} \delta \\ V \\ \lambda \end{bmatrix} = \begin{bmatrix} -0.2060 \\ 0.9637 \\ 0.2000 \end{bmatrix}$$

where λ is increased by the stepsize $\sigma = 0.1$ pu.

Correcting the values yields the second update:

$$\begin{bmatrix} \delta \\ V \\ \lambda \end{bmatrix} = \begin{bmatrix} -0.2105 \\ 0.9570 \\ 0.2000 \end{bmatrix}$$

Iterations 3 and 4

The third and fourth iterations progress similarly. The values to this point are summarized:

λ	V	δ	σ
0.1000	0.9847	-0.1017	0.1000
0.2000	0.9570	-0.2105	0.1000
0.3000	0.9113	-0.3354	0.1000
0.4000	0.8268	-0.5050	0.1000

Beyond this point, the loadflow fails to converge for a stepsize of $\sigma = 0.1$. The method is nearing the point of maximum power flow (the tip of the nose curve) as indicated by the rapid decline in voltage for relatively small changes in λ. At this point, the continuation parameter is switched from λ to V to ensure that the corrector step will converge. The predictor step is modified such that:

$$\begin{bmatrix} d\delta \\ dV \\ d\lambda \end{bmatrix} = \begin{bmatrix} d\delta_0 \\ dV_0 \\ d\lambda_0 \end{bmatrix} + \sigma \begin{bmatrix} [J_{LF}] & -\lambda \\ & 0 \\ 0 & -1 & 0 \end{bmatrix}^{-1} \begin{bmatrix} 0 \\ 0 \\ 1 \end{bmatrix}$$

where the -1 in the last row (the e_k vector) now corresponds to V rather than λ. The minus sign indicates that the predictor step will reduce the voltage magnitude by the stepsize σ. The stepsize σ is reduced to 0.025pu, which is a value more appropriate for changes in voltage magnitude.

The corrector step is also modified when the continuation parameter switches to voltage magnitude. The new augmented equations become:

$$0 = f_1(\delta, V, \lambda) = -\lambda - 0.995V\left(\cos\left(\delta - 95.7°\right) - V\cos(84.3°)\right) \quad (3.122)$$
$$0 = f_2(\delta, V, \lambda) = -0.995V\sin\left(\delta - 95.7°\right) - 0.995V^2\sin(84.3°) \quad (3.123)$$
$$0 = f_3(\delta, V, \lambda) = V - V^{\text{predicted}} \quad (3.124)$$

which cannot be solved with a traditional powerflow program due to the last equation. This equation is necessary to keep the Newton-Raphson iteration nonsingular. Fortunately, the Newton-Raphson iteration uses the same iteration matrix as the predictor matrix:

$$\begin{bmatrix} [J_{LF}] & -\lambda \\ & 0 \\ 0 & -1 & 0 \end{bmatrix}^{-1} \begin{bmatrix} 0 \\ 0 \\ 1 \end{bmatrix} \left(\begin{bmatrix} \delta \\ V \\ \lambda \end{bmatrix}^{(k+1)} - \begin{bmatrix} \delta \\ V \\ \lambda \end{bmatrix}^{(k)} \right) = - \begin{bmatrix} f_1 \\ f_2 \\ f_3 \end{bmatrix} \quad (3.125)$$

thus minimizing the computational requirement.

Note that the corrector step is now a horizontal correction in voltage. The voltage magnitude is held constant while λ and δ are corrected. These iterates proceed as

	Predicted				Corrected	
λ	V	δ	σ	λ	V	δ
0.4196	0.8019	-0.5487	0.0250	0.4174	0.8019	-0.5474
0.4326	0.7769	-0.5887	0.0250	0.4307	0.7769	-0.5876
0.4422	0.7519	-0.6268	0.0250	0.4405	0.7519	-0.6260
0.4487	0.7269	-0.6635	0.0250	0.4472	0.7269	-0.6627
0.4525	0.7019	-0.6987	0.0250	0.4511	0.7019	-0.6981
0.4538	0.6769	-0.7328	0.0250	0.4525	0.6769	-0.7323
0.4528	0.6519	-0.7659	0.0250	0.4517	0.6519	-0.7654

Note that at the last updates, the load parameter λ has started to decrease with decreasing voltage. This indicates that the continuation power flow is now starting to map out the lower half of the nose curve. However, while the iterations are still close to the tip of the nose curve, the Jacobian will still be ill conditioned, so it is a good idea to take several more steps before switching the continuation parameter from voltage magnitude back to λ.

	Predicted				Corrected	
λ	V	δ	σ	λ	V	δ
0.4497	0.6269	-0.7981	0.0250	0.4487	0.6269	-0.7977
0.4447	0.6019	-0.8295	0.0250	0.4438	0.6019	-0.8291
0.4380	0.5769	-0.8602	0.0250	0.4371	0.5769	-0.8598
0.4296	0.5519	-0.8902	0.0250	0.4288	0.5519	-0.8899
0.4197	0.5269	-0.9197	0.0250	0.4190	0.5269	-0.9194

After switching the continuation parameter back to λ, the e_k vector becomes

$$e_k = [0 \quad 0 \quad -1]$$

where the -1 indicates that the continuation parameter λ will be decreasing (i.e., the power is decreasing back to the base case). The predictor/corrector steps proceed as above yielding

	Predicted				Corrected	
λ	V	δ	σ	λ	V	δ
0.3190	0.2899	-1.1964	0.1000	0.3190	0.3564	-1.1088
0.2190	0.2187	-1.2554	0.1000	0.2190	0.2317	-1.2387
0.1190	0.1165	-1.3565	0.1000	0.1190	0.1220	-1.3496
0.0190	0.0166	-1.4553	0.1000	0.0190	0.0191	-1.4523

These values are combined in the PV curve shown in Figure 3.12. Note the change in step size when the continuation parameter switches from λ to voltage near the tip of the PV curve. The appropriate choice of step size is problem dependent and can be adaptively changed to improve computational efficiency. ■

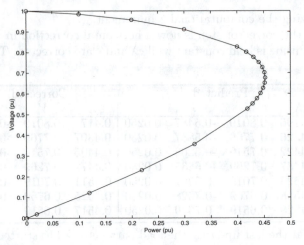

FIGURE 3.12
PV curve for system of Example 3.10

3.3.6   Three-phase Power Flow

Another special purpose power flow application is for three-phase power flow analysis. Although much of the power system analysis is performed on balanced three phase systems using one line equivalent diagrams, certain situations call for a three phase analysis. In particular, in the situation where the transmission system is not balanced due to non-transposed lines or when the loads are considerably unbalanced, it may be desirable to perform a full three phase load flow to ascertain the effect on the individual line flows and bus voltages. The development of the three phase load flow is similar to that of the single phase equivalent, except that the mutual coupling between lines must be incorporated into the admittance matrix, yielding a $3n \times 3n$ matrix with elements Y_{ij}^{pq} where the subscript ij indicates bus number $(1 \leq (i,j) \leq n)$ and the superscript pq indicates phase $(p,q \in [a,b,c])$.

The incorporation of each phase individually leads to the similar, but slightly more complex, three-phase load flow equations:

$$0 = \Delta P_i^p = P_i^{inj,p} - V_i^p \sum_{q \in (a,b,c)} \sum_{j=1}^{N_{bus}} V_j^q Y_{ij}^{pq} \cos\left(\theta_i^p - \theta_j^q - \phi_{ij}^{pq}\right) \quad (3.126)$$

$$0 = \Delta Q_i^p = Q_i^{inj,p} - V_i^p \sum_{q \in (a,b,c)} \sum_{j=1}^{N_{bus}} V_j^q Y_{ij}^{pq} \sin\left(\theta_i^p - \theta_j^q - \phi_{ij}^{pq}\right) \quad (3.127)$$

$$i = 1, \ldots, N_{bus} \text{ and } p \in (a,b,c)$$

There are three times as many load flow equations as in the single phase equivalent load flow equations. Generator (PV) buses are handled similarly with the following exceptions:

1. $\theta^a = 0°, \theta^b = -120°, \theta^c = 120°$ for the swing bus

2. All generator voltage magnitudes and active powers in each phase must be equal since generators are designed to produce balanced output

A three-phase load flow "flat start" is to set each bus voltage magnitude to

$$V_i^a = 1.0\angle 0°$$
$$V_i^b = 1.0\angle - 120°$$
$$V_i^c = 1.0\angle 120°$$

The system Jacobian used in the Newton-Raphson solution of the load flow equations will have a possible $(3(2n) \times 3(2n))$ or $36n^2$ entries. The Jacobian partial derivatives are found in the same manner as with the single phase equivalent system except that the derivatives must also be taken with respect to phase differences. For example,

$$\frac{\partial \Delta P_i^a}{\partial \theta_j^b} = V_i^a V_j^b Y_{ij}^{ab} \sin\left(\theta_i^a - \theta_j^b - \phi_{ij}^{ab}\right) \tag{3.128}$$

which is similar to the single phase equivalent system. Similarly

$$\frac{\partial \Delta P_i^a}{\partial \theta_i^a} = -V_i^a \sum_{q \in (a,b,c)} \sum_{j=1}^{N_{bus}} V_j^q Y_{ij}^{pq} \sin\left(\theta_i^p - \theta_j^q - \phi_{ij}^{pq}\right) + (V_i^a)^2 Y_{ii}^{pp} \cos\left(\phi_{ii}^{pp}\right) \tag{3.129}$$

The remaining partial derivatives can be calculated in a similar manner and the solution process of the three-phase power flow follows the method outlined in Section 3.3.1.

3.4 Problems

1. Prove that the Newton-Raphson iteration will diverge for the following functions regardless of choice of initial condition

   (a) $f(x) = x^2 + 1$

   (b) $f(x) = 7x^4 + 3x^2 + \pi$

2. Devise an iteration formula for computing the fifth root of any positive real number.

3. Solve

$$0 = 4y^2 + 4y + 52x - 19$$
$$0 = 169x^2 + 3y^2 + 111x - 10y - 10$$

4. Solve

$$0 = x - 2y + y^2 + y^3 - 4$$
$$0 = -xy + 2y^2 - 1$$

5. Write a *generalized* (for any system) power flow program. Your program should:

   (a) Read in load, voltage, and generation data. You may assume that bus #1 corresponds to the swing bus.

   (b) Read in line and transformer data and create the Y_{bus} matrix.

   (c) Solve the power flow equations using the Newton-Raphson algorithm, for a stopping criterion of $\|f(x^k)\| \le \epsilon = 0.0005$.

   (d) Calculate all dependent unknowns, line flows, and line losses.

   The Newton-Raphson portion of the program should call the lufact and permute subroutines. Your program should give you the option of using either a "flat start" or "previous values" as an initial guess. The easiest way to accomplish this is to read and write to the same data file. Note that the first run must be a "flat start" case.

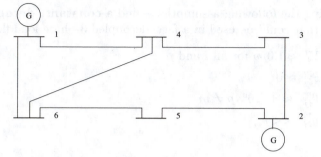

FIGURE 3.13
Ward-Hale 6 bus system

6. The data for the system shown in Figure 3.13 are given below:

No.	Type	$|V|$	θ	P_{gen}	Q_{gen}	P_{load}	Q_{load}
1	0	1.05	0	0	0	0.25	0.1
2	1	1.05	0	0.5	0	0.15	0.05
3	2	1.00	0	0	0	0.275	0.11
4	2	1.00	0	0	0	0	0
5	2	1.00	0	0	0	0.15	0.09
6	2	1.00	0	0	0	0.25	0.15

No.	To	From	R	X	B
1	1	4	0.020	0.185	0.009
2	1	6	0.031	0.259	0.010
3	2	3	0.006	0.025	0.000
4	2	5	0.071	0.320	0.015
5	4	6	0.024	0.204	0.010
6	3	4	0.075	0.067	0.000
7	5	6	0.025	0.150	0.017

Calculate the load flow solution for the system data given above. Remember to calculate all dependent unknowns, line-flows, and line losses.

7. Modify your loadflow program so that you are using a *decoupled load flow* (i.e., assume $\left[\frac{\partial \Delta P}{\partial V}\right] = 0$ and $\left[\frac{\partial \Delta Q}{\partial \theta}\right] = 0$). Repeat problem 6. Discuss the difference in convergence between the decoupled and the full Newton-Raphson Power Flows.

8. Increase the line resistances by 75% (i.e. multiply all resistances by 1.75) and repeat problem 6 and problem 7. Discuss your findings.

9. Using a continuation power flow, map out the "PV" curve for the original system data by increasing/decreasing the load on bus 6 holding a constant P/Q ratio from $P = 0$ to the point of maximum power transfer.

10. Making the following assumptions, find a **constant, decoupled** Jacobian that could be used in a fast, decoupled 3-phase load flow.

- $V_i^p \approx 1.0 pu$ for all i and p
- $\theta_{ij}^{pp} \approx 0$
- $\theta_{ij}^{pm} \approx \pm 120° \quad p \neq m$
- $g_{ij}^{pm} << b_{ij}^{pm}$

Chapter 4

Sparse Matrix Solution Techniques

A sparse matrix is one that has "very few" non-zero elements. A sparse system is one in which its mathematical description gives rise to sparse matrices. Any large system that can be described by coupled nodes may give rise to a sparse matrix if the majority of nodes in the system have very few connections. Many systems in engineering and science result in sparse matrix descriptions. Large systems in which each node is connected to only a handful of other nodes include the mesh points in a finite-element-analysis, nodes in an electronic circuit, and the busbars in an electric power network. For example, power networks may contain thousands of nodes (busbars), but the average connectivity of electric power network nodes is three; each node is connected, on average, to three other nodes. This means that in a system comprised of a thousand nodes, the non-zero percentage of the descriptive system matrix is

$$\frac{\frac{4 \text{ non-zeros elements}}{\text{row}} \times 1000 \text{ rows}}{1000 \times 1000 \text{ elements}} \times 100\% = 0.4\% \text{ non-zero elements}$$

Thus, if only the non-zero elements were stored in memory, they would require only 0.4% of the memory requirements of the full 1000×1000 matrix. Full storage of an $n \times n$ system matrix grows as n^2, whereas the sparse storage of the same system matrix increases only linearly as $\sim n$. Thus significant storage and computational savings can be realized by exploiting sparse storage and solution techniques. Another motivating factor in exploiting sparse matrix solution techniques is the computational effort involved in solving matrices with large percentages of zero elements. Consider the solution of the linear problem

$$Ax = b$$

where A is sparse. The factorization of L and U from A requires a significant number of multiplications where one or both of the factors may be zero. If it is known ahead of time where the zero elements reside in the matrix, these multiplications can be avoided (since their product will be zero) and significant computational effort can be saved. The salient point here is that these computations are skipped altogether. A person performing an LU factorization by hand can note which values are zero and skip those particular multiplications. A computer, however, does not have the ability to "see" the zero elements. Therefore the sparse solution technique must be formulated in

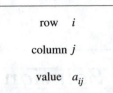

FIGURE 4.1
Basic storage element for a_{ij}

such a way as to avoid zero computations altogether and operate only upon non-zero elements.

In this chapter, both the storage and computational aspects of sparse matrix solutions will be explored. Several storage techniques will be discussed and ordering techniques to minimize computational effort will be developed.

4.1   Storage Methods

In sparse storage methods, only the non-zero elements of the $n \times n$ matrix A are stored in memory, along with the indexing information needed to transverse the matrix from element to element. Thus each element must be stored with its real value (a_{ij}) and its position indices (row and column) in the matrix. The basic storage unit may be visualized as the object shown in Figure 4.1.

In addition to the basic information, indexing information must also be included in the object, such as a link to the next element in the row, or the next element in the column, as shown in Figure 4.2.

The only additional information necessary to fully transverse the matrix either by row or column is an indication of the first element in each row or column. This is a stand-alone set of links that point to the first element in each row or column.

Example 4.1
Determine a linked list representation for the sparse matrix:

$$A = \begin{bmatrix} -1 & 0 & -2 & 0 & 0 \\ 2 & 8 & 0 & 1 & 0 \\ 0 & 0 & 3 & 0 & -2 \\ 0 & -3 & 2 & 0 & 0 \\ 1 & 2 & 0 & 0 & -4 \end{bmatrix}$$

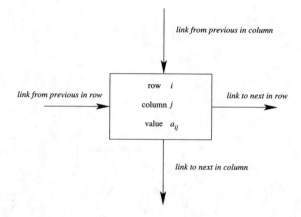

FIGURE 4.2
Storage element for element a_{ij} with links

Solution 4.1 A linked list representation of this matrix is shown in Figure 4.3. The last element of each column and row are linked to a null point. Note that each object is linked to its adjacent neighbors in the matrix, both by column and by row. In this way, the entire matrix can be transversed in any direction by starting at the first element and following the links until the desired element is reached. ■

If a command requires a particular matrix element, then by either choosing the column or row, the element can be located by progressing along the links. If during the search, the null point is reached, then the desired element does not exist and a value of zero is returned. Additionally, if the matrix elements are linked by increasing index and an element is reached that has a greater index than the desired element, then the progression terminates and a value of zero is returned.

A linked list representation for a sparse matrix is not unique and the elements do not necessarily have to be linked in increasing order by index. However, ordering the elements by increasing index leads to a simplified search since the progression can be terminated before reaching the null point if the index of the linked object exceeds the desired element. If the elements are not linked in order, the entire row or column must be transversed to determine whether or not the desired element is non-zero. The drawback to an ordered list is that the addition of new non-zero elements to the matrix requires the update of both row and column links.

Example 4.2
Insert the matrix element $A(4, 5) = 10$ to the linked list of Example 4.1.

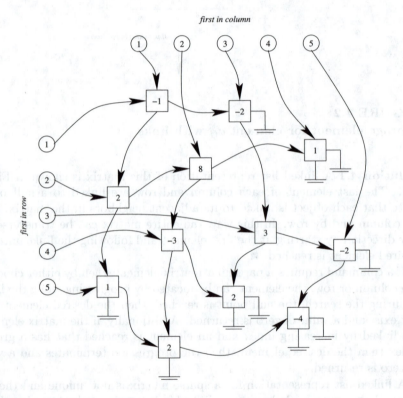

FIGURE 4.3
Linked list for Example 4.1

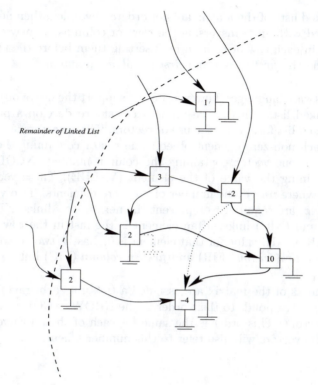

FIGURE 4.4
Insertion of matrix element $A(4,5) = 10$

Solution 4.2 The insertion of the new element is shown in Figure 4.4. The addition of this element requires two transversals of the matrix to insert the new element and to update the links; one transversal by row and one by column. Starting at the *first in row* link for row 4 (value=-3), the elements are progressed by link. Following the links and monitoring the column indices, it is discovered that the element with column index 3 (value=2) is the last element in the row since it points to null. Since the new element has column index 5, it is inserted between the (value=2) element and the null point in the row linked list. Similarly, starting at the *first in column* link for column 5 (value=-2), the column is transversed and inserted between the elements with row indices 3 (value=-2) and 5 (value=-4). The column links are updated to reflect the insertion. ∎

If the linked lists of the matrix are not ordered by index, then new elements can be added without transversing the rows or columns. A new element can be inserted in each row or column by inserting them before the first element and updating the *first in row* and *first in column* pointers.

Many software languages, however, do not support the use of objects, pointers, and linked lists. In this case it is necessary to develop a procedure to mimic a linked list format by the use of vectors. Three vectors are required to represent each non-zero element object: one vector containing the row number (NROW), one vector containing the column number (NCOL), and one vector containing the value of the element (VALUE). These vectors are of length nnz where nnz is the number of non-zero elements. Two vectors, also of length nnz, are required to represent the next-in-row links (NIR) and the next-in-column (NIC) links. If an element is the last in the row or column, then the NIR or NIC value for that element is 0. Lastly, two vectors of length n contain the *first in row* (FIR) and *first in column* (FIC) links.

The elements of the matrix are assigned a (possibly arbitrary) numbering scheme that corresponds to their order in the NROW, NCOL, VALUE, NIR, and NIC vectors. This order is the same for each of these five vectors. The FIR and FIC vectors will also refer to this number scheme.

Example 4.3

Find the vectors NROW, NCOL, VALUE, NIR, NIC, FIR, and FIC for the sparse matrix of Example 4.1.

Solution 4.3 The matrix of Example 4.1 is reproduced below with the numbering scheme given in parentheses to the left of each non-zero element. The numbering scheme is sequential by row and goes from 1 to $nnz = 12$.

$$A = \begin{bmatrix} (1)\ -1 & 0 & (2)\ -2 & 0 & 0 \\ (3)\ 2 & (4)\ 8 & 0 & (5)\ 1 & 0 \\ 0 & 0 & (6)\ 3 & 0 & (7)\ -2 \\ 0 & (8)\ -3 & (9)\ 2 & 0 & 0 \\ (10)\ 1 & (11)\ 2 & 0 & 0 & (12)\ -4 \end{bmatrix}$$

The ordering scheme indicated yields the following nnz vectors:

k	VALUE	NROW	NCOL	NIR	NIC
1	-1	1	1	2	3
2	-2	1	3	0	6
3	2	2	1	4	10
4	8	2	2	5	8
5	1	2	4	0	0
6	3	3	3	7	9
7	-2	3	5	0	12
8	-3	4	2	9	11
9	2	4	3	0	0
10	1	5	1	11	0
11	2	5	2	12	0
12	-4	5	5	0	0

and the following n vectors:

	FIR	FIC
1	1	1
2	3	4
3	6	2
4	8	5
5	10	7

Consider the matrix element $A(2,2) = 8$. It is the 4th element in the numbering scheme, so its information is stored in the fourth place in vectors VALUE, NROW, NCOL, NIR, and NIC. Thus VALUE(4)=8, NROW(4)=2, and NCOL(4)=2. The next element in row 2 is $A(2,4) = 1$ and it is element 5 in the numbering scheme. Therefore NROW(4)=5, signifying that element 5 follows element 4 in its row (note however that it does not indicate which row they are in). Similarly, the next element in column 2 is $A(4,2) = -3$ and it is element 8 in the numbering scheme. Therefore NCOL(4)=8. ■

4.2 Sparse Matrix Representation

Sparse matrices arise as the result of the mathematical modeling of a sparse system. In many cases, the system has a naturally occurring physical network representation or lends itself to a physically intuitive representation. In these cases, it is often informative to visualize the connectivity of the system by graphical means. In the graphical representation, each node of the graph corresponds to a node in the system. Each edge of the graph corresponds to a branch of the network. As with a network, the graph, consisting of vertices and edges, is often represented by a set of points in the plane joined by a line representing each edge. Matrices that arise from the mathematical model of a graphically represented network are structurally symmetric. In other words, if the matrix element a_{ij} is non-zero, then the matrix element a_{ji} is

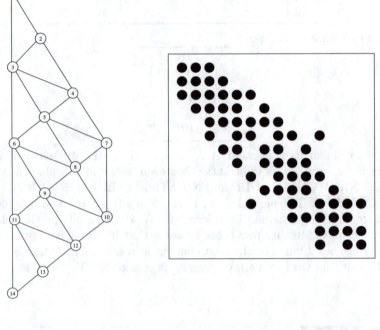

(a)

(b)

FIGURE 4.5
(a) A finite element grid model, (b) The corresponding matrix

also non-zero. This implies that if node i is connected to node j, then node j is also connected to node i. Matrices that are not structurally symmetric can by made so by adding an element of value zero in the appropriate position within the matrix.

In addition to a graphical representation, it is also common to visualize sparse matrices by a matrix that is clear except for an identifying symbol (such as a $\times, \bullet, *$, or other mark) to represent the position of the non-zero elements in the matrix. The finite element grid of the trapezoid shown in Figure 4.5(a) gives rise to the sparse matrix structure shown in Figure 4.5(b). Note that the ordering of the matrix is not unique; another numbering scheme for the nodes will result in an alternate matrix structure.

4.3 Ordering Schemes

Node ordering schemes are important in minimizing the number of multiplications and divisions required for both L and U triangularization and forward/backward substitution. A good ordering will result in the addition of few *fills* to the triangular factors during the LU factorization process. A *fill* is a non-zero element in the L or U matrix that was zero in the original A matrix. If A is a full matrix, $\alpha = \frac{n^3 - n}{3}$ multiplications and divisions are required for the LU factorization process and $\beta = n^2$ multiplications and divisions are required for the forward/backward substitution process. The number of multiplications and divisions required can be substantially reduced in sparse matrix solutions if a proper node ordering is used.

Example 4.4
Determine the number of multiplications, divisions, and fills required for the solution of the system shown in Figure 4.6.

Solution 4.4 The LU factorization steps yield

$$q_{11} = a_{11}$$
$$q_{21} = a_{21}$$
$$q_{31} = a_{31}$$
$$q_{41} = a_{41}$$
$$q_{51} = a_{51}$$

$$q_{12} = a_{12}/q_{11}$$
$$q_{13} = a_{13}/q_{11}$$

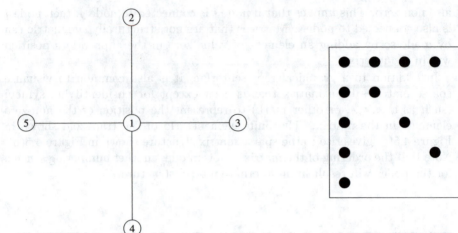

FIGURE 4.6
Graph and matrix for Example 4.4

$$q_{14} = a_{14}/q_{11}$$
$$q_{15} = a_{15}/q_{11}$$

$$q_{22} = a_{22} - q_{21}q_{12}$$
$$q_{32} = a_{32} - q_{31}q_{12}$$
$$q_{42} = a_{42} - q_{41}q_{12}$$
$$q_{52} = a_{52} - q_{51}q_{12}$$

$$q_{23} = (a_{23} - q_{21}q_{13})/q_{22}$$
$$q_{24} = (a_{24} - q_{21}q_{14})/q_{22}$$
$$q_{25} = (a_{25} - q_{21}q_{15})/q_{22}$$

$$q_{33} = a_{33} - q_{31}q_{13} - q_{32}q_{23}$$
$$q_{43} = a_{43} - q_{41}q_{13} - q_{42}q_{23}$$
$$q_{53} = a_{53} - q_{51}q_{13} - q_{52}q_{23}$$

$$q_{34} = (a_{34} - q_{31}q_{14} - q_{32}q_{24})/q_{33}$$
$$q_{35} = (a_{35} - q_{31}q_{15} - q_{32}q_{25})/q_{33}$$

$$q_{44} = a_{44} - q_{41}q_{14} - q_{42}q_{24} - q_{43}q_{34}$$
$$q_{54} = a_{54} - q_{51}q_{14} - q_{52}q_{24} - q_{53}q_{34}$$

$$q_{45} = (a_{45} - q_{41}q_{15} - q_{22}q_{25} - q_{43}q_{35})/q_{44}$$

$$q_{55} = a_{55} - q_{51}q_{15} - q_{52}q_{25} - q_{53}q_{35} - q_{54}q_{45}$$

The multiplications and divisions required for the LU factorization are summarized by row and column.

row	column	multiplications	divisions	fills
1		0	0	
	1	0	4	
2		4	0	a_{32}, a_{42}, a_{52}
	2	3	3	a_{23}, a_{24}, a_{25}
3		6	0	a_{43}, a_{53}
	3	4	2	a_{34}, a_{35}
4		6	0	a_{54}
	4	3	1	a_{45}
5		4	0	

Therefore $\alpha = 40$ is the total number of multiplications and divisions in the LU factorization. The forward $(Ly = b)$ and backward $(Ux = y)$ substitution steps yield:

$$y_1 = b_1/q_{11}$$
$$y_2 = (b_2 - q_{21}y_1)/q_{22}$$
$$y_3 = (b_3 - q_{31}y_1 - q_{32}y_2)/q_{33}$$
$$y_4 = (b_4 - q_{41}y_1 - q_{42}y_2 - q_{43}y_3)/q_{44}$$
$$y_5 = (b_5 - q_{51}y_1 - q_{52}y_2 - q_{53}y_3 - q_{54}y_4)/q_{55}$$
$$x_5 = y_5$$
$$x_4 = y_4 - q_{45}x_5$$
$$x_3 = y_3 - q_{35}x_5 - q_{34}x_4$$
$$x_2 = y_2 - q_{25}x_5 - q_{24}x_4 - q_{23}x_3$$
$$x_1 = y_1 - q_{15}x_5 - q_{14}x_4 - q_{13}x_3 - q_{12}x_2$$

| row | forward | | backward | |
	multiplications	divisions	multiplications	divisions
1	0	1	4	0
2	1	1	3	0
3	2	1	2	0
4	3	1	1	0
5	4	1	0	0

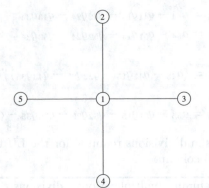

FIGURE 4.7
Graph for Example 4.4

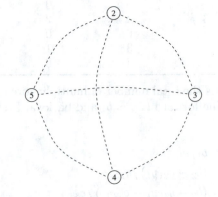

FIGURE 4.8
Resulting fills after removing node 1

Thus $\beta = 25$ is the total number of multiplications and divisions in the forward and backward substitution steps. The total number of multiplications and divisions for the solution of $Ax = b$ is $\alpha + \beta = 36$. ∎

A fill occurs when a matrix element that was originally zero becomes non-zero during the factorization process. This can be visually simulated using a graphical approach. Consider the graph of Example 4.4 shown again in Figure 4.7.

In this numbering scheme, the row and column corresponding to node 1 is factorized first. This corresponds to the removal of node 1 from the graph. When node 1 is removed, all of the vertices to which it was connected must then be joined. Each edge added represents two fills in the Q matrix (q_{ij} and q_{ji}) since Q is symmetric. The graph after the removal of node 1 is

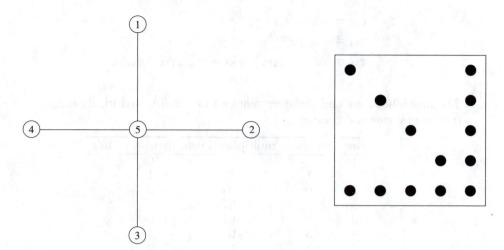

FIGURE 4.9
Graph and matrix for Example 4.5

shown in Figure 4.8. The dashed lines indicate that six fills will occur as a result: q_{23}, q_{24}, q_{25}, q_{34}, q_{35}, and q_{45}. These are the six fills that are also listed in the solution of the example.

Example 4.5
Determine the number of multiplications, divisions, and fills required for the solution of the system shown in Figure 4.9.

Solution 4.4 The LU factorization steps yield

$$q_{11} = a_{11}$$
$$q_{51} = a_{51}$$
$$q_{15} = a_{15}/q_{11}$$
$$q_{22} = a_{22}$$
$$q_{25} = a_{25}/q_{22}$$
$$q_{52} = a_{52}$$
$$q_{33} = a_{33}$$
$$q_{53} = a_{53}$$
$$q_{35} = a_{35}/q_{33}$$

$$q_{44} = a_{44}$$

$$q_{54} = a_{54}$$
$$q_{45} = a_{45}/q_{44}$$
$$q_{55} = a_{55} - q_{51}q_{15} - q_{52}q_{25} - q_{53}q_{35} - q_{54}q_{45}$$

The multiplications and divisions required for the LU factorization are summarized by row and column.

row	column	multiplications	divisions	fills
1		0	0	
	1	0	1	
2		0	0	
	2	0	1	
3		0	0	
	3	0	1	
4		0	0	
	4	0	1	
5		4	0	

Therefore $\alpha = 8$ is the total number of multiplications and divisions in the LU factorization. The forward $(Ly = b)$ and backward $(Ux = y)$ substitution steps yield:

$$y_1 = b_1/q_{11}$$
$$y_2 = b_2/q_{22}$$
$$y_3 = b_3/q_{33}$$
$$y_4 = b_4/q_{44}$$
$$y_5 = (b_5 - q_{51}y_1 - q_{52}y_2 - q_{53}y_3 - q_{54}y_4)/q_{55}$$
$$x_5 = y_5$$
$$x_4 = y_4 - q_{45}x_5$$
$$x_3 = y_3 - q_{35}x_5$$
$$x_2 = y_2 - q_{25}x_5$$
$$x_1 = y_1 - q_{15}x_5$$

| | forward | | backward | |
row	multiplications	divisions	multiplications	divisions
1	0	1	1	0
2	0	1	1	0
3	0	1	1	0
4	0	1	1	0
5	4	1	0	0

Thus $\beta = 13$ is the total number of multiplications and divisions in the forward and backward substitution steps. The total number of multiplications and divisions for the solution of $Ax = b$ is $\alpha + \beta = 21$. ∎

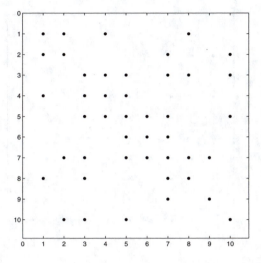

FIGURE 4.10
Matrix for Example 4.6

Even though both original matrices had the same number of non-zero elements, there is a significant reduction in the number of multiplications and divisions by simply renumbering the vertices of the matrix graph. This is due, in part, by the number of fills that occurred during the LU factorization of the matrix. The Q matrix of Example 4.4 became full, whereas the Q matrix of Example 4.5 retained the same sparse structure as the original A matrix. From these two examples, it can be concluded that although various node orders do not affect the accuracy of the linear solution, the ordering scheme greatly affects the time in which the solution is achieved. A good ordering scheme is one in which the resulting Q matrix has a similar structure to the original A matrix. This means that the number of fills is minimized. This objective forms the basis for a variety of ordering schemes. The problem of optimal ordering is an NP-complete problem [41], but several schemes have been developed that provide near-optimal results.

Example 4.6
Determine number of fills, α, and β for the matrix shown in Figure 4.10 as currently ordered.

Solution 4.6 The first step is to determine where the fills from LU factorization will occur. By observation, the fills will occur in the places designated by the $\triangle$ in the matrix shown in Figure 4.11. From the figure, the number of fills is 24.

Rather than calculating the number of multiplications and divisions required for LU factorization and forward/backward substitution, there is a

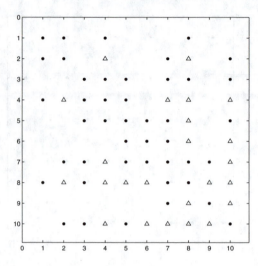

FIGURE 4.11
Matrix with fills for Example 4.6

handy way of calculating α and β directly from the filled matrix.

$$\alpha = \sum_{i=1}^{n} (\text{nnz in column } i \text{ below } q_{ii} + 1) \times (\text{nnz in row } i \text{ to right of } q_{ii}) \quad (4.1)$$

$$\beta = nnz \text{ of matrix } Q \quad (4.2)$$

Using equations (4.1) and (4.2),

$$\alpha = (3 \times 4) + (4 \times 5) + (5 \times 6) + (4 \times 5) + (4 \times 5) + (3 \times 4)$$
$$+ (3 \times 4) + (2 \times 3) + (1 \times 2) + (0 \times 1) = 134$$

and $\beta = nnz = 68$ for the Q matrix shown in Figure 4.11, thus $\alpha + \beta = 202$. Compare this with the $\alpha + \beta = 430$. ∎

Even without an ordering scheme, the sparse matrix solution process yields over a 50% reduction in computation. One goal of an ordering scheme is to introduce the least number of fills in the factored matrix Q to minimize the number of multiplications and divisions α. A second goal is also to minimize β, which is the number of multiplications and divisions in the forward/backward substitution step. These dual objectives lead to several approaches to ordering.

4.3.1 Scheme 0

From Examples 4.4 and 4.5, it can be generalized that a better ordering is achieved if the nodes are ordered into a lower-right pointing "arrow" matrix. One rapid method of achieving this effect is to number the nodes according to their degree, where the degree of a node is defined as the number of edges connected to it. In this scheme, the nodes are ordered from lowest degree to highest degree.

Scheme 0

1. Calculate the degree of all vertices.

2. Choose the node with the lowest degree. Place in ordering scheme.

3. In case of a tie, choose node with lowest natural ordering.

4. Return to step 2.

Example 4.7
Using Scheme 0, reorder the matrix of Example 4.6. Calculate α, β and the number of fills for this ordering.

Solution 4.7 The degrees of each of the nodes are given below:

node	degree
1	3
2	3
3	5
4	3
5	5
6	2
7	6
8	3
9	1
10	3

Applying Scheme 0, the new ordering is

$$\text{Ordering } 0 = [9\ 6\ 1\ 2\ 4\ 8\ 10\ 3\ 5\ 7]$$

Reordering the matrix of Example 4.6 to reflect this ordering yields the matrix (with fills) shown in Figure 4.12. Note how the non-zeros elements begin to resemble the desired lower-right pointing arrow. The Scheme 0 ordering results in 16 fills as compared to 24 with the original ordering. From the matrix and equations (4.1) and (4.2), $\alpha = 110$ and $\beta = 60$, thus $\alpha + \beta = 170$ which is a considerable reduction over the original $\alpha + \beta = 202$. ∎

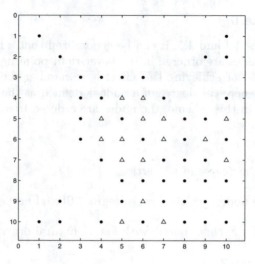

FIGURE 4.12
Matrix with fills for Example 4.7

4.3.2 Scheme I

Scheme 0 offers simplicity and speed of generation, but does not directly take into account the effect of fills on the ordering procedure. To do this, the effect of eliminating the nodes as they are ordered must be taken into account. This modification is given in Scheme I.

Scheme I

1. Calculate the degree of all vertices.

2. Choose the node with the lowest degree. Place in ordering scheme. Eliminate it and update degrees accordingly.

3. In case of a tie, choose node with lowest natural ordering.

4. Return to step 1.

Scheme I is also known by many names including the Markowitz algorithm [23], the Tinney I algorithm [37], or most generally as the minimum degree algorithm.

Example 4.8
Using Scheme I, reorder the matrix of Example 4.6. Calculate α, β and the number of fills for this ordering.

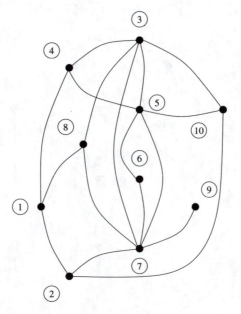

FIGURE 4.13
Graph of the matrix in Figure 4.10

Solution 4.8 The ordering for Scheme I takes into account the effect of fills on the ordering as nodes are placed in the ordering scheme and eliminated. This algorithm is best visualized using the graphical representation of the matrix. The graph of the orginal unordered matrix of Figure 4.10 is shown in Figure 4.13.

The degrees of each of the nodes are given below:

node	degree
1	3
2	3
3	5
4	3
5	5
6	2
7	6
8	3
9	1
10	3

From the degrees, the node with the lowest degree is ordered first. Node 9 has the lowest degree with only one connection. Its elimination does not cause any fills. The updated graph is shown in Figure 4.14.

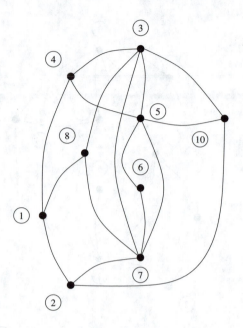

FIGURE 4.14
Updated graph with the removal of node 9

The updated degree of each of the nodes is given below:

node	degree
1	3
2	3
3	5
4	3
5	5
6	2
7	5
8	3
10	3

Node 7 now has one less degree. Applying the Scheme I algorithm again indicates that the next node to be chosen is node 6, with a degree of 2. Node 6 is connected to both node 5 and node 7. Since there is a pre-existing connection between these nodes, the elimination of node 6 does not create a fill between nodes 5 and 6. The elimination of node 6 is shown in Figure 4.15.

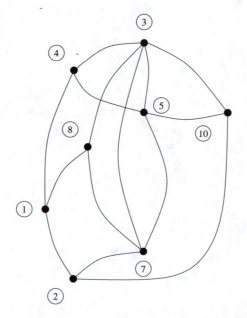

FIGURE 4.15
Updated graph with the removal of node 6

The new node degrees are

node	degree
1	3
2	3
3	5
4	3
5	4
7	4
8	3
10	3

As a result of the elimination of node 6, the degrees of nodes 5 and 7 decrease by one. Applying the Scheme I algorithm again indicates that the nodes with the lowest degrees are nodes [1 2 4 8 10]. Since there is a tie between these nodes, the node with the lowest natural ordering, node 1, is chosen and eliminated. Node 1 is connected to nodes 2, 4, and 8. None of these nodes is connected; therefore, the elimination of node 1 creates three fills: 4–8, 4–2, and 2–8. These fills are shown with the dashed edges in Figure 4.16.

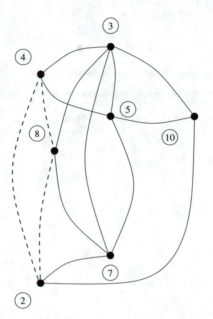

FIGURE 4.16
Updated graph with the removal of node 1

The new node degrees after the removal of node 1 are

node	degree
2	4
3	5
4	4
5	5
7	5
8	4
10	3

The addition of the three fills increased the degrees of nodes 2, 4, and 8. Applying the Scheme I algorithm again indicates that the node with the lowest degree is node 10. There is no tie in degree this time. Node 10 is chosen and eliminated. The elimination of node 10 creates two fills between nodes 2–5 and 2–3. These fills are shown with the dashed edges in Figure 4.17.

Continuing to apply the Scheme I algorithm successively until all nodes have been chosen and eliminated yields the following final ordering:

$$\text{Ordering I} = [9 \ 6 \ 1 \ 10 \ 4 \ 2 \ 3 \ 5 \ 7 \ 8]$$

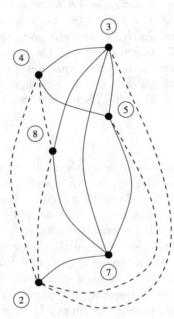

FIGURE 4.17
Updated graph with the removal of node 10

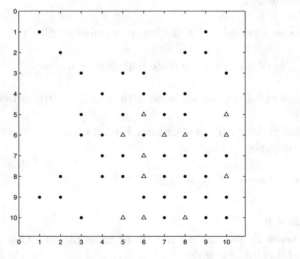

FIGURE 4.18
Matrix with fills for Example 4.8

Reordering the matrix of Example 4.8 to reflect this ordering yields the matrix (with fills) shown in Figure 4.18. Note how the non-zero elements continue to resemble the desired lower-right pointing arrow. The Scheme I ordering results in 12 fills as compared to 24 with the original ordering, and 16 with Scheme 0. From the matrix and equations (4.1) and (4.2), $\alpha = 92$ and $\beta = 56$, thus $\alpha + \beta = 148$ which is a considerable reduction over the original $\alpha + \beta = 202$, and the $\alpha + \beta = 170$ of Scheme 0. ∎

4.3.3   Scheme II

Scheme 0 offers a rapid way to order the nodes to give a quick "once-over" and obtain a reasonable ordering. It requires little computation beyond calculating the degrees of each node of the matrix. Scheme I takes this approach one step further. It still relies on the minimum-degree approach, but it includes a simulation of the LU factorization process to update the node degrees at each step of the factorization. One further improvement to this approach is to develop a scheme that endeavors to minimize the number of fills at each step of the factorization. Thus, at each step, each elimination alternative is considered and the number of resulting fills is calculated. This scheme is also known as the Berry algorithm and the Tinney II algorithm and is summarized below:

Scheme II

1. For each node, calculate the number of fills that would result from its elimination.

2. Choose the node with the lowest number of fills.

3. In case of a tie, choose node with lowest degree.

4. In case of a tie, choose node with lowest natural ordering.

5. Place node in ordering scheme. Eliminate it and update fills and degrees accordingly.

6. Return to step 1.

Example 4.9
Using Scheme II, reorder the matrix of Example 4.6. Calculate α, β and the number of fills for this ordering.

Solution 4.9 The ordering for Scheme II takes into account the effect of fills on the ordering as nodes are placed in the ordering scheme and eliminated.

The degrees and resulting fills are given below:

node	degree	fills if eliminated	edges created
1	3	3	2–4, 2–8, 4–8
2	3	3	1–7, 1–10, 7–10
3	5	6	4–7, 4–8, 4–10, 5–8, 7–10, 8–10
4	3	2	1–3, 1–5
5	5	6	3–6, 4–6, 4–7, 4–10, 6–10, 7–10
6	2	0	none
7	6	12	2–3, 2–5, 2–6, 2–8, 2–9, 3–6, 3–9, 5–8, 5–9, 6–8, 6–9, 8–9
8	3	2	1–3, 1–7
9	1	0	none
10	3	2	2–3, 2–5

From this list, the elimination of nodes 6 or 9 will not result in any additional edges, or fills. Since there is a tie, the node with the lowest degree is chosen. Thus, node 9 is chosen and eliminated. The number of fills and degrees is updated to apply the Scheme II algorithm again.

node	degree	fills if eliminated	edges created
1	3	3	2–4, 2–8, 4–8
2	3	3	1–7, 1–10, 7–10
3	5	6	4–7, 4–8, 4–10, 5–8, 7–10, 8–10
4	3	2	1–3, 1–5
5	5	6	3–6, 4–6, 4–7, 4–10, 6–10, 7–10
6	2	0	none
7	5	7	2–3, 2–5, 2–6, 2–8, 3–6, 5–8, 6–8
8	3	2	1–3, 1–7
10	3	2	2–3, 2–5

The next node to be eliminated is node 6 because it creates the fewest fills if eliminated. This node is therefore chosen and eliminated. The number of fills and degrees is again updated.

node	degree	fills if eliminated	edges created
1	3	3	2–4, 2–8, 4–8
2	3	3	1–7, 1–10, 7–10
3	5	6	4–7, 4–8, 4–10, 5–8, 7–10, 8–10
4	3	2	1–3, 1–5
5	5	6	3–6, 4–6, 4–7, 4–10, 6–10, 7–10
7	5	7	2–3, 2–5, 2–6, 2–8, 3–6, 5–8, 6–8
8	3	2	1–3, 1–7
10	3	2	2–3, 2–5

The two nodes that create the fewest fills are nodes 4 and 8. Both nodes have the same number of degrees; therefore, the node with the lowest natural ordering, node 4, is chosen and eliminated.

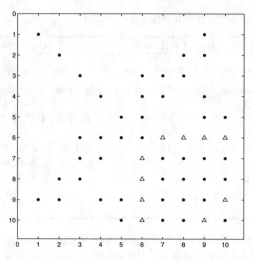

FIGURE 4.19
Matrix with fills for Example 4.9

The Scheme II algorithm continues until all nodes have been added to the ordering scheme and subsequently eliminated. Scheme II results in the following ordering:

$$\text{Ordering II} = [9\ 6\ 4\ 8\ 2\ 1\ 3\ 5\ 7\ 10]$$

Reordering the matrix of Example 4.6 to reflect the ordering of Scheme II yields the ordering with fills shown in Figure 4.19. This ordering yields only 10 fills, leading to an $\alpha = 84, \beta = 54$, and $\alpha + \beta = 138$. This represents a computational effort of only 68% of the original unordered system. ∎

Scheme I endeavors to reduce the number of multiplications and divisions in the LU factorization process, whereas Scheme II focuses on reducing the multiplications and divisions in the forward/backward substitution process. Scheme 0 offers simplicity and speed of generation, but the performance improvement of Scheme I offsets the additional algorithm complexity [37]. Scheme II, however, frequently does not offer enough of an improvement to merit implementation. The decision of which scheme to implement is problem dependent and is best left up to the user.

4.3.4   Other Schemes

Modifications to these algorithms have been introduced to reduce computational requirements. These modifications are summarized below [12]. The first modification to the minimum-degree algorithm is the use of mass elimination, inspired by the concept of indistinguishable nodes [11]. This modification allows a subset of nodes to be eliminated at one time. If two nodes x and y satisfy

$$Adj(y) \cup \{y\} = Adj(x) \cup \{x\} \tag{4.3}$$

where *Adj(y)* indicates the set of nodes adjacent to y, then nodes x and y are said to be indistinguishable and can be numbered consecutively in the ordering. This also reduces the number of nodes to be considered in an ordering, since only a representative node from each set of indistinguishable nodes needs to be considered. This accelerates the degree update step of the minimum-degree algorithm, which is typically the most computationally intensive step. Using mass elimination, the degree update is required only for the representative nodes.

The idea of incomplete degree update allows avoiding degree update for nodes that are known not to be minimum degree. Between two nodes u and v, node v is said to be outmatched by u if [7]

$$Adj(u) \cup \{u\} \subseteq Adj(v) \cup \{v\} \tag{4.4}$$

Thus, if a node v becomes outmatched by u in the elimination process, the node u can be eliminated before v in the minimum-degree ordering algorithm. From this, it follows that it is not necessary to update the degree of v until node u has been eliminated. This further reduces the time-consuming degree update steps.

Another modification to the minimum-degree algorithm is one in which all possible nodes of minimum degree are eliminated before the degree update step. At a given step in the elimination process, the elimination of node y does not change the structure of the remaining nodes not in *Adj(y)*. The multiple-minimum-degree (MMD) algorithm delays degree update of the nodes in *Adj(y)*, and chooses another node with the same degree as y to eliminate. This process continues until there are no more nodes left with the same degree as y. This algorithm was found to perform as well as the minimum-degree algorithm regarding the number of fills introduced [22]. In addition, it was found that the MMD algorithm performed faster. This was attributed to the identification of indistinguishable and outmatched nodes earlier in the algorithm, as well as the reduced number of degree updates.

Ties often occur for a given criteria (degrees or fills) in an ordering algorithm. The tie breaker often falls back on the natural ordering of the original matrix. It has been recognized that the natural ordering greatly affects the factorization in terms of number of fills and computation time. Thus it is often preferable to use a rapid "pre-conditioning" ordering before applying the ordering algorithm. Scheme 0 offers one such pre-ordering, but to date no consistent optimum method for pre-ordering has been developed that works well for all types of problems.

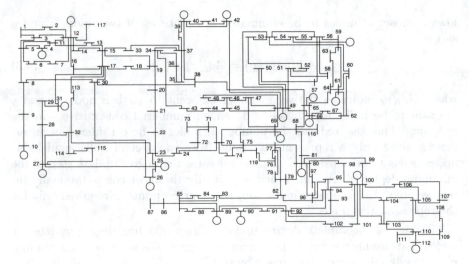

FIGURE 4.20
IEEE 118 bus system

4.4 Power System Applications

Large sparse matrices occur frequently in power system applications, including state estimation, power flow analysis, and transient and dynamic stability simulations. Computational efficiency of these applications depends heavily on their formulation and the use of sparse matrix techniques. To better understand the impact of sparsity on power system problems, consider the power flow Jacobian of the IEEE 118 bus system shown in Figure 4.20.

The Jacobian of this system has 1051 non-zero elements and has the structure shown in Figure 4.21(a). Note the dominance of the main diagonal and then the two sub-diagonals which result from the $\frac{\partial \Delta Q}{\partial \delta}$ and $\frac{\partial \Delta P}{\partial V}$ sub-Jacobians. The LU factorization of this Jacobian yields the structure shown in Figure 4.21(b). This matrix has 14849 non-zero elements. Notice that the two sub-diagonals have created a large number of fills extending between them and the main diagonal.

Figure 4.22(a) shows the structure of the load flow Jacobian reordered according to Scheme 0. In this reordering, the presence of the sub-diagonals is gone. The LU factorization of the Scheme 0 reordered Jacobian yields the structure shown in Figure 4.22(b). This matrix has only 1869 non-zero elements, which is almost an order of magnitude reduction from the non-ordered Jacobian.

Figure 4.23(a) shows the structure of the load flow Jacobian reordered according to Scheme 1. Note how the elements are gradually pulling in to the

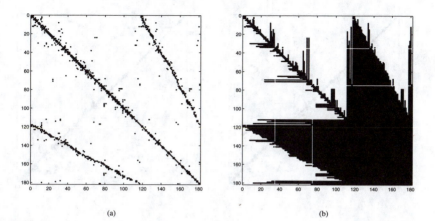

FIGURE 4.21
IEEE 118 bus system (a) Jacobian, and (b) LU factors

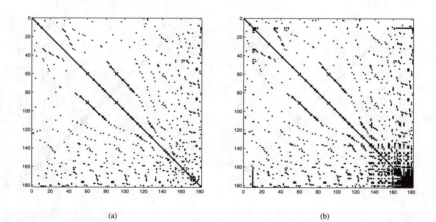

FIGURE 4.22
IEEE 118 bus system Scheme 0 (a) Jacobian, and (b) LU factors

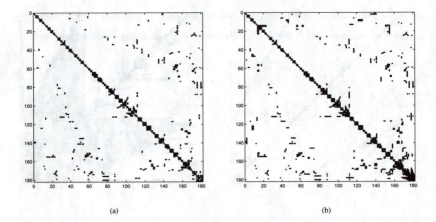

(a)   (b)

FIGURE 4.23
IEEE 118 bus system Scheme 1 (a) Jacobian, and (b) LU factors

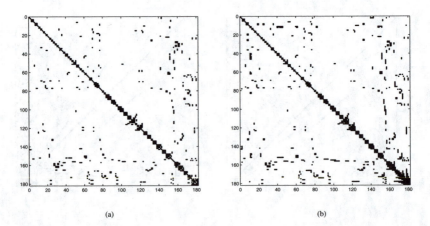

(a)   (b)

FIGURE 4.24
IEEE 118 bus system Scheme 2 (a) Jacobian, and (b) LU factors

main diagonal, which leads to a decrease in the number of fills. The LU factorization of the Scheme 1 ordering is shown in Figure 4.23(b), which has 1455 non-zero elements.

Lastly, Figure 4.24(a) shows the structure of the Scheme 2 reordered Jacobian which yields the LU factorization in Figure 4.24(b). This ordering yields only 1421 non-zero elements, which is more than a full order of magnitude reduction. The LU factorization solution time for a sparse matrix is on the order of n^2 multiplications and divisions. The non-reordered load flow solution would require on the order of 220.5×10^6 multiplications and divisions per iteration, whereas the Scheme 2 reordered load flow solution would require only 2.02×10^6 multiplications and divisions. Thus, the solution of the reordered system is over 100 times faster than the original system! When the solution time is multiplied by the number of iterations in a Newton-Raphson power flow or by the number of time steps in a time-domain integration, it would be computationally foolhardy to not use a reordering scheme.

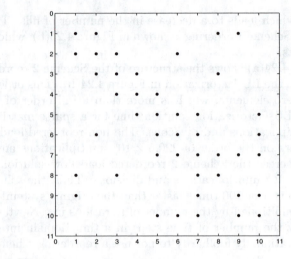

FIGURE 4.25
Sparse Test System

4.5   Problems

1. Verify equations (4.1) and (4.2) for calculating α and β.

2. Let A and B be two sparse (square) matrices of the same dimension. How can the graph of $C = A + B$ be characterized with respect to the graphs of A and B?

3. Consider the matrix

$$
A = \begin{bmatrix}
* & * & & & * & \\
* & * & * & & & * \\
& * & * & & & \\
& & & * & * & \\
* & & & * & * & * \\
& * & & & * & *
\end{bmatrix}
$$

   (a) Draw the graphical representation of A. How many multiplications and divisions will the LU factorization of A require?

   (b) Reorder the matrix using the new node numbering $\phi = [1, 3, 4, 2, 5, 6]$. Draw the graphical representation of the reordered matrix. How many multiplications and divisions will the LU factorization of the reordered matrix require?

4. For the matrix shown in Figure 4.25:

   (a) Using the given ordering, compute $\alpha + \beta$.

(b) Reorder the nodes in the network using Scheme 0 discussed in class. Compute $\alpha + \beta$ for this ordering.

(c) Reorder the nodes in the network using Scheme I. Compute $\alpha + \beta$ for this ordering.

(d) Reorder the nodes in the network using Scheme II. Compute $\alpha + \beta$ for this ordering.

5. Write a subroutine *sparmat* for sparse matrix storage that will

   - Read in data line by line in the format:

   $$i \quad j \quad a_{ij}$$

   where the end of the data is signified by a 0 in the first column

   - Sequentially build the vectors FIR, FIC, NIR, NIC, NROW, NCOL, and Value as defined in class. **Do not explicitly create the matrix** A.

6. Write a subroutine *sparvec* for sparse vector storage that will

   - Read in data line by line in the format:

   $$i \quad b_i$$

   where the end of the data is signified by a 0 in the first column

   - Sequentially build the vectors index, next, and Value. **Do not explicitly create the vector** b.

7. For the data given below, use *sparmat* and *sparvec* to create the sparse storage vectors.

\multicolumn{3}{c}{A matrix}			b vector	
i	j	a_{ij}	i	b_i
7	10	2.0	2	5
2	6	1.5	9	2
9	1	4.7	3	-1
5	5	-18.5		
8	7	2.8		
1	1	-15.0		
4	3	3.8		
6	7	6.1		
8	3	3.3		
5	7	4.4		
10	6	2.5		
6	5	1.1		
3	2	5.2		
7	8	2.9		
9	9	-12.1		
3	4	3.0		
7	6	5.6		
10	9	4.7		
8	8	-10.8		
1	9	4.5		
7	5	3.9		
5	6	7.2		
9	10	4.9		
5	4	0.8		
8	1	3.4		
5	10	4.5		
2	3	5.0		
6	6	-9.8		
7	9	1.8		
4	5	0.7		
7	7	-21.2		
1	2	4.4		
10	5	5.4		
3	8	3.1		
9	7	1.6		
4	4	-5.1		
6	10	2.7		
10	10	-16.9		
2	1	4.7		
3	3	-17.7		
1	8	3.5		
10	7	2.1		
2	2	-13.0		
6	2	1.2		

8. Write a subroutine *sparLU* that modifies your LU factorization routine to incorporate the sparse storage vectors of Problem 2 and apply it to the data of Problem 4 to compute the sparse LU factors (in sparse vector form).

9. Write a subroutine *sparsub* that modifies your forward/backward substitution routine *sub* to incorporate the sparse storage vectors of Problem 2 and apply it to the data of Problem 4 to solve the sparse linear system

$$Ax = b$$

10. Write a subroutine *scheme0* that will input the sparse vectors FIR, FIC, NIR, NIC, NROW, NCOL, and Value and will output the same vectors reordered according to Scheme 0, and calculate $\alpha + \beta$.

11. Write a subroutine *scheme1* that will input the sparse vectors FIR, FIC, NIR, NIC, NROW, NCOL, and Value and will output the same vectors reordered according to Scheme I, and calculate $\alpha + \beta$.

12. Write a subroutine *scheme2* that will input the sparse vectors FIR, FIC, NIR, NIC, NROW, NCOL, and Value and will output the same vectors reordered according to Scheme II, and calculate $\alpha + \beta$.

Chapter 5

Numerical Integration

Dynamic systems may frequently be modeled by systems of *ordinary differential equations* (or ODEs) of the form

$$\dot{x}(t) = f(x,t) \quad x(t_0) = x_0 \tag{5.1}$$

where $x(t) \in R^n$ is a time-varying function that depends on the initial condition x_0. Such problems are often referred to as "initial value problems." A system of nonlinear differential equations cannot typically be solved analytically. In other words, a closed form expression for $x(t)$ cannot be found directly from equation (5.1), but rather must be solved numerically.

In the numerical solution of equation (5.1), a sequence of points $x_0, x_1, x_2 \ldots$, are computed that approximate the true solution at a set of time points $t_0, t_1, t_2, \ldots$. The time interval between adjacent time points is called the *time step* and an integration algorithm advances the numerical solution by one time step with each application. The time step $h_{n+1} = t_{n+1} - t_n$ may be constant for all time intervals over the entire integration interval $t \in [t_0, t_N]$, or may vary at each step.

The basic integration algorithm advances the solution from t_n to t_{n+1} with integration step size h_{n+1} based on a calculation that involves previously computed values $x_n, x_{n-1}, \ldots$ and functions $f(x_n, t_n), f(x_{n-1}, t_{n-1}), \ldots$. Each practical integration algorithm must satisfy certain criteria concerning

1. numerical accuracy,

2. numerical stability, and

3. numerical efficiency.

Numerical accuracy ensures that the numerical error incurred at each step of the integration remains bounded. The global error of an integration error is the total error accrued over a given time interval. The global error at time t_n is given by

$$\text{global error} = \|x(t_n) - x_n\|$$

where $x(t_n)$ is the exact solution to equation (5.1) at time t_n and x_n is the approximated solution. Of course, it is impossible to determine the global error exactly if the solution $x(t)$ is not known analytically, but it is possible to establish bounds on the error incurred at each step of the integration method.

The numerical stability of an integration algorithm implies that errors incurred at each step do not propagate to future times. Numerical efficiency is a function of the amount of computation required at each time step and the size of the step sizes between adjacent time intervals. Each of these criteria will be discussed in greater detail later in this chapter after an introduction to several different forms of integration algorithms.

5.1 One-Step Methods

The basic form of an integration algorithm is one that advances the solution from x_n to x_{n+1} using only the information currently available. This type of solution is called a *one-step* method, in that only information from one step of the integration algorithm is used. The family of one-step methods has the advantage of conserving memory, since only the previous solution must be retained. Several well-known methods fall into this category.

5.1.1 Taylor Series-based methods

One important class of integration methods is derived from using the Taylor series expansion of equation (5.1). Let $\hat{x}(t)$ denote the exact solution to equation (5.1). Expanding $\hat{x}(t)$ in a Taylor series about $t = t_n$ and evaluating the series at $t = t_{n+1}$ yields the following series expansion for $\hat{x}(t_{n+1})$:

$$\hat{x}(t_{n+1}) = \hat{x}(t_n) + \dot{x}(t_n)(t_{n+1} - t_n)$$
$$+ \frac{1}{2}\ddot{x}(t_n)(t_{n+1} - t_n)^2 + \ldots + \frac{1}{p!}x^{(p)}(t_n)(t_{n+1} - t_n)^p + h.o.t.$$

where *h.o.t.* stands for *higher order terms* of the expansion. If the time step $h = t_{n+1} - t_n$ then

$$\hat{x}(t_{n+1}) = \hat{x}(t_n) + h\dot{x}(t_n) + \frac{h^2}{2}\ddot{x}(t_n) + \ldots + \frac{h^p}{p!}x^{(p)}(t_n) + h.o.t.$$

From equation (5.1), $\dot{x}(t) = f(x, t)$, therefore

$$\hat{x}(t_{n+1}) - h.o.t. = \hat{x}(t_n) + hf(x(t_n), t_n)$$
$$+ \frac{h^2}{2}f'(x(t_n), t_n) + \ldots + \frac{h^p}{p!}f^{(p-1)}(x(t_n), t_n) \quad (5.2)$$

If the higher order terms are small, then a good approximation x_{n+1} to $\hat{x}(t_{n+1})$ is given by the right hand side of equation (5.2).

In general, the Taylor series-based integration methods can be expressed as

$$x_{n+1} = x_n + hT_p(x_n) \tag{5.3}$$

where

$$T_p(x_n) = f(x(t_n), t_n) + \frac{h^2}{2} f'(x(t_n), t_n) + \ldots + \frac{h^p}{p!} f^{(p-1)}(x(t_n), t_n)$$

and the integer p is called the *order* of the integration method. This method is very accurate for large p, but is not computationally efficient for large p since it requires a large number of function derivatives and evaluations.

5.1.2 Forward-Euler method

For $p = 1$, the Taylor series-based integration algorithm is given by:

$$x_{n+1} = x_n - hf(x_n, t_n) \tag{5.4}$$

which is also the well-known *Euler* or *forward Euler* method.

5.1.3 Runge-Kutta methods

A second order Taylor's method can be derived for $p = 2$.

$$\begin{aligned}
x_{n+1} &= x_n + hT_2(x_n, t_n) \\
&= x_n + hf(x_n, t_n) + \frac{h^2}{2} f'(x_n, t_n) \\
&= x_n + hf(x_n, t_n) + \frac{h^2}{2} \left[\frac{\partial f}{\partial x}(x_n, t_n) + \frac{\partial f}{\partial t}(x_n, t_n) \right]
\end{aligned}$$

As the order of the Taylor's method increases, so does the number of derivatives and partial derivatives. In many cases, the analytic derivation of the derivatives can be replaced by a numerical approximation. One of the most commonly known higher-order Taylor series-based integration methods is the Runge-Kutta method, where the derivatives are replaced by approximations. The fourth-order Runge-Kutta method is given by

$$x_{n+1} = x_n + hK_4(x_n, t_n) \tag{5.5}$$

where K_4 is an approximation to T_4:

$$\begin{aligned}
K_4 &= \frac{1}{6} [k_1 + 2k_2 + 2k_3 + k_4] \\
k_1 &= f(x_n, t_n) \\
k_2 &= f\left(x_n + \frac{h}{2}k_1, t_n + \frac{h}{2}\right) \\
k_3 &= f\left(x_n + \frac{h}{2}k_2, t_n + \frac{h}{2}\right) \\
k_4 &= f(x_n + hk_3, t_n + h)
\end{aligned}$$

where each k_i represents the slope (derivative) of the function at four different points. The slopes are then weighted $\left[\frac{1}{6} \ \frac{2}{6} \ \frac{2}{6} \ \frac{1}{6}\right]$ to approximate the T_4 function.

The advantages of Taylor series-based methods is that the method is straightforward to program and only depends on the previous time step. These methods (especially the Runge-Kutta methods) suffer from difficult error analysis, however, since the derivatives are approximated and not found analytically. Therefore the integration step size is typically chosen conservatively (small), and computational efficiency may be lost.

5.2 Multistep Methods

Another approach to approximating the solution $x(t)$ of equation (5.1) is to approximate the nonlinear function as a polynomial of degree k such that

$$\hat{x}(t) = \alpha_0 + \alpha_1 t + \alpha_2 t^2 + \ldots + \alpha_k t^k \tag{5.6}$$

where the coefficients $\alpha_0, \alpha_1, \ldots, \alpha_k$ are constant. It can be proven that any function can be approximated arbitrarily closely (within a pre-determined ε) with a polynomial of sufficiently high degree on a finite interval $[t_0, t_N]$. The polynomial approximation can be related to the solution of equation (5.1) through the introduction of *multistep* methods. A multistep method is one in which the approximation x_{n+1} can be a function of any number of previous numerical approximations $x_n, x_{n-1}, \ldots$ and corresponding functions $f(x_n, t_n), f(x_{n-1}, t_{n-1}), \ldots$ unlike one-step methods (such as the Runge-Kutta) which depend only on the information from the immediately previous step. In general,

$$x_{n+1} = a_0 x_n + a_1 x_{n-1} + \ldots + a_p x_{n-p} + h\left[b_{-1} f(x_{n+1}, t_{n+1}) + b_0 f(x_n, t_n)\right.$$
$$\left. + b_1 f(x_{n-1}, t_{n-1}) + \ldots + b_p f(x_{n-p}, t_{n-p})\right] \tag{5.7}$$

To relate the integration method to the polynomial approximation, a relationship between the coefficients must be determined. A k-degree polynomial is uniquely determined by $k + 1$ coefficients $(\alpha_0, \ldots, \alpha_k)$. The numerical integration method has $2p + 3$ coefficients; therefore, the coefficients must be chosen such that

$$2p + 3 \geq k + 1 \tag{5.8}$$

The order of the numerical integration method is the highest degree k of a polynomial in t for which the numerical solution coincides with the exact solution. The coefficients may be determined by selecting a set of linear basis functions $[\phi_1(t) \ \phi_2(t), \ \ldots, \ \phi_k(t)]$ such that

$$\phi_j(t) = t^j \quad j = 0, 1, \ldots, k$$

and solving the set of multistep equations

$$\phi_j\left(t_{n+1}\right) = \sum_{i=0}^{p} a_i \phi_j\left(t_{n-i}\right) + h_{n+1}\left[\sum_{i=-1}^{p} b_i \dot{\phi}\left(t_{n-i}\right)\right]$$

for all $j = 0, 1, \ldots, k$.

This method can be applied to derive several first order numerical integration methods. Consider the case where $p = 0$ and $k = 1$. This satisfies the constraint of equation (5.8); thus, it is possible to determine multistep coefficients that will result in an exact polynomial of degree 1. The set of basis functions for $k = 1$ is

$$\phi_0(t) = 1 \tag{5.9}$$
$$\phi_1(t) = t \tag{5.10}$$

which lead to the derivatives

$$\dot{\phi}_0(t) = 0 \tag{5.11}$$
$$\dot{\phi}_1(t) = 1 \tag{5.12}$$

and the multistep equation

$$x_{n+1} = a_0 x_n + b_{-1} h_{n+1} f\left(x_{n+1}, t_{n+1}\right) + b_0 h_{n+1} f\left(x_n, t_n\right) \tag{5.13}$$

Representing the multistep method of equation (5.13) in terms of basis functions yields the following two equations

$$\phi_0\left(t_{n+1}\right) = a_0 \phi_0(t_n) + b_1 h_{n+1} \dot{\phi}_0\left(t_{n+1}\right) + h_{n+1} b_0 \dot{\phi}_0\left(t_n\right) \tag{5.14}$$
$$\phi_1\left(t_{n+1}\right) = a_0 \phi_1(t_n) + b_1 h_{n+1} \dot{\phi}_1\left(t_{n+1}\right) + h_{n+1} b_0 \dot{\phi}_1\left(t_n\right) \tag{5.15}$$

Substituting the choice of basis functions of equations (5.9) and (5.10) into equations (5.14) and (5.15) results in

$$1 = a_0(1) + b_{-1} h_{n+1}(0) + h_{n+1} b_0(0) \tag{5.16}$$
$$t_{n+1} = a_0 t_n + b_{-1} h_{n+1}(1) + b_0 h_{n+1}(1) \tag{5.17}$$

From equation (5.16), the coefficient $a_0 = 1$. Recalling that $t_{n+1} - t_n = h_{n+1}$, equation (5.17) yields

$$b_{-1} + b_0 = 1 \tag{5.18}$$

This choice of order and degree leads to two equations in three unknowns; therefore, one of them may be chosen arbitrarily. By choosing $a_0 = 1, b_{-1} = 0$, and $b_0 = 1$, Euler's method is once again obtained:

$$x_{n+1} = x_n + h_{n+1} f\left(x_n, t_n\right)$$

However, if $a_0 = 1, b_{-1} = 1$, and $b_0 = 0$, a different integration method is obtained:

$$x_{n+1} = x_n + h_{n+1} f(x_{n+1}, t_{n+1}) \tag{5.19}$$

This particular integration method is frequently called the *backward Euler* method. Note that in this method, the coefficient b_{-1} is not identically zero; thus, the expression for x_{n+1} depends implicitly on the function $f(x_{n+1}, t_{n+1})$. Methods in which $b_{-1} \neq 0$ are referred to as *implicit* methods; otherwise they are *explicit*. Since there is an implicit (and often nonlinear) dependence on x_{n+1}, implicit integration methods must usually be solved iteratively at each time interval.

Consider now the case where $p = 0$, and $k = 2$. In this case, $2p + 3 = k + 1$ and the coefficients can be uniquely determined. Choosing the basis functions as previously with $\phi_2(t) = t^2$ and $\dot{\phi}_2(t) = 2t$ yields the following three equations:

$$1 = a_0(1) + b_{-1} h_{n+1}(0) + h_{n+1} b_0(0) \tag{5.20}$$
$$t_{n+1} = a_0 t_n + b_{-1} h_{n+1}(1) + b_0 h_{n+1}(1) \tag{5.21}$$
$$t_{n+1}^2 = a_0 t_n^2 + h_{n+1} (b_{-1}(2t_{n+1}) + b_0(2t_n)) \tag{5.22}$$

If $t_n = 0$, then $t_{n+1} = h_{n+1}$, and equations (5.20) through (5.22) yield $a_0 = 1, b_{-1} = \frac{1}{2}$, and $b_0 = \frac{1}{2}$, thus

$$x_{n+1} = x_n + \frac{1}{2} h_{n+1} [f(x_{n+1}, t_{n+1}) + f(x_n, t_n)] \tag{5.23}$$

This second-order integration method is called the *trapezoidal* method and it is also implicit. This formula is called the trapezoidal method since the second term of equation (5.23) can be interpreted as being the area under a trapezoid. The trapezoidal is considered a two step method since information from both t_n and t_{n+1} is required.

Example 5.1
Numerically solve

$$\ddot{x}(t) = -x(t) \quad x(0) = 1 \tag{5.24}$$

using the Euler, backward Euler, trapezoidal, and Runge-Kutta methods for different fixed step sizes.

Solution 5.1 This second-order differential equation must first be converted to ODE format by defining $x_1 = x$ and $x_2 = \dot{x}$. Then

$$\dot{x}_1 = x_2 = f_1(x_1, x_2) \quad x_1(0) = 1 \tag{5.25}$$
$$\dot{x}_2 = -x_1 = f_2(x_1, x_2) \tag{5.26}$$

By inspection, the analytic solution to this set of equations is

$$x_1(t) = \cos t \qquad (5.27)$$
$$x_2(t) = -\sin t \qquad (5.28)$$

Typically it is not possible to find the exact solution, but in this example, the exact solution will be used to compare the numerical solutions against.

Forward-Euler
Applying the forward-Euler method to the ODEs yields

$$x_{1,n+1} = x_{1,n} + h f_1(x_{1,n}, x_{2,n}) \qquad (5.29)$$
$$= x_{1,n} + h x_{2,n} \qquad (5.30)$$
$$x_{2,n+1} = x_{2,n} + h f_2(x_{1,n}, x_{2,n}) \qquad (5.31)$$
$$= x_{2,n} - h x_{1,n} \qquad (5.32)$$

or in matrix form:

$$\begin{bmatrix} x_{1,n+1} \\ x_{2,n+1} \end{bmatrix} = \begin{bmatrix} 1 & h \\ -h & 1 \end{bmatrix} \begin{bmatrix} x_{1,n} \\ x_{2,n} \end{bmatrix} \qquad (5.33)$$

Backward-Euler
Applying the backward-Euler method to the ODEs yields

$$x_{1,n+1} = x_{1,n} + h f_1(x_{1,n+1}, x_{2,n+1}) \qquad (5.34)$$
$$= x_{1,n} + h x_{2,n+1} \qquad (5.35)$$
$$x_{2,n+1} = x_{2,n} + h f_2(x_{1,n+1}, x_{2,n+1}) \qquad (5.36)$$
$$= x_{2,n} - h x_{1,n+1} \qquad (5.37)$$

or in matrix form:

$$\begin{bmatrix} x_{1,n+1} \\ x_{2,n+1} \end{bmatrix} = \begin{bmatrix} 1 & -h \\ h & 1 \end{bmatrix}^{-1} \begin{bmatrix} x_{1,n} \\ x_{2,n} \end{bmatrix} \qquad (5.38)$$

In the solution of equation (5.38), the inverse of the matrix is not found explicitly, but rather the equations would be solved using LU factorization.

Trapezoidal
Applying the trapezoidal method to the ODEs yields

$$x_{1,n+1} = x_{1,n} + \frac{1}{2} h \left[f_1(x_{1,n}, x_{2,n}) + f_1(x_{1,n+1}, x_{2,n+1}) \right] \qquad (5.39)$$

$$= x_{1,n} + \frac{1}{2} h \left[x_{2,n} + x_{2,n+1} \right] \qquad (5.40)$$

$$x_{2,n+1} = x_{2,n} + \frac{1}{2} h \left[f_2(x_{1,n}, x_{2,n}) + f_2(x_{1,n+1}, x_{2,n+1}) \right] \qquad (5.41)$$

$$= x_{2,n} - \frac{1}{2} h \left[x_{1,n} + x_{1,n+1} \right] \qquad (5.42)$$

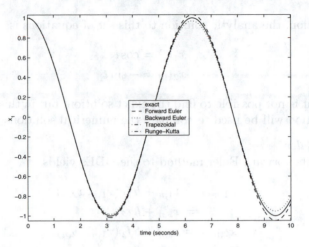

FIGURE 5.1
Numerical solutions for Example 5.1

or in matrix form:

$$\begin{bmatrix} x_{1,n+1} \\ x_{2,n+1} \end{bmatrix} = \begin{bmatrix} 1 & -\frac{1}{2}h \\ \frac{1}{2}h & 1 \end{bmatrix}^{-1} \begin{bmatrix} 1 & \frac{1}{2}h \\ -\frac{1}{2}h & 1 \end{bmatrix} \begin{bmatrix} x_{1,n} \\ x_{2,n} \end{bmatrix} \tag{5.43}$$

Runge-Kutta
Applying the Runge-Kutta method to the ODEs yields

$$\begin{aligned}
k_{11} &= x_{2,n} & k_{21} &= -x_{1,n} \\
k_{12} &= x_{2,n} + \frac{h}{2}k_{11} & k_{22} &= -x_{1,n} - \frac{h}{2}k_{21} \\
k_{13} &= x_{2,n} + \frac{h}{2}k_{12} & k_{23} &= -x_{1,n} - \frac{h}{2}k_{22} \\
k_{14} &= x_{2,n} + hk_{13} & k_{24} &= -x_{1,n} - hk_{23}
\end{aligned}$$

and

$$x_{1,n+1} = x_{1,n} + \frac{h}{6}\left(k_{11} + 2k_{12} + 2k_{13} + k_{14}\right) \tag{5.44}$$

$$x_{2,n+1} = x_{2,n} + \frac{h}{6}\left(k_{21} + 2k_{22} + 2k_{23} + k_{24}\right) \tag{5.45}$$

The numerical solution of equation (5.24) for each of the methods is shown in Figure 5.1 including the exact solution $\cos t$. Note that the trapezoidal and Runge-Kutta methods are nearly indistinguishable from the exact solution. Since the forward- and backward-Euler methods are first-order methods, they are not as accurate as the higher order trapezoidal and Runge-Kutta methods. Note that the forward-Euler method generates a numerical solution whose magnitude is slightly larger than the exact solution and is increasing with time. Conversely, the backward-Euler method generates a numerical solution

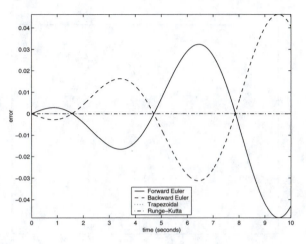

FIGURE 5.2
Error in numerical solutions for Example 5.1

whose magnitude is slightly less than the exact solution and is decreasing with time. Both properties are due to the *local truncation error* of the methods. The forward-Euler method has a tendency to generate numerical solutions that increase with time (under-damped), whereas the backward-Euler method tends to add damping to the numerical solution. Therefore, caution must be used when using either of these first-order methods for numerical integration.

Figure 5.2 shows the global error with time for each of the numerical methods. Note that the errors for the forward and backward-Euler methods are equal in magnitude, but opposite in sign. This relationship will be further discussed later in this chapter. The numerical errors for the trapezoidal and the Runge-Kutta methods are reproduced in Figure 5.3 on a magnified scale. The errors in each method are comparable even though the trapezoidal method is a second order polynomial approximation method and the Runge-Kutta is a fourth order Taylor method. Section 5.3 will further explore the development of expressions for estimating the error of various integration methods. ■

When implicit methods, such as the trapezoidal method, are used to solve nonlinear systems of differential equations, the system of equations must be solved iteratively at each time step. For example, consider the following nonlinear system of equations:

$$\dot{x} = f(x(t), t) \quad x_0 = x(t_0) \qquad (5.46)$$

Applying the trapezoidal method to numerically integrate this system results in the following discretized system:

$$x_{n+1} = x_n + \frac{h}{2} \left[f\left(x_n, t_n\right) + f\left(x_{n+1}, t_{n+1}\right) \right] \qquad (5.47)$$

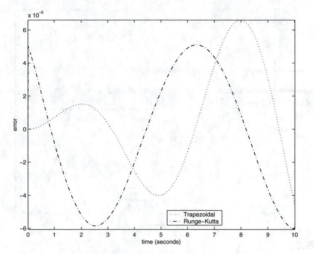

FIGURE 5.3
Error in trapezoidal and Runge-Kutta numerical solutions for Example 5.1

Since this nonlinear expression is implicit in x_{n+1}, it must be solved numerically:

$$x_{n+1}^{k+1} = x_{n+1}^k - \left[I - \frac{h}{2}\frac{\partial f}{\partial x} \right]^{-1}\Bigg|_{x_{n+1}^k} \left(x_{n+1}^k - x_n - \frac{h}{2}\left[f\left(x_n\right) + f\left(x_{n+1}^k\right) \right] \right)$$

(5.48)

where k is the Newton-Raphson iteration index, I is the identity matrix, and x_n is the converged value from the previous time step.

5.2.1 Adam's methods

Recall that the general class of multistep methods may be represented by

$$x_{n+1} = \sum_{i=0}^{p} a_i x_{n-i} + h \sum_{i=-1}^{p} b_i f\left(x_{n-i}, t_{n-i}\right)$$

(5.49)

A numerical multistep algorithm will give the exact value for x_{n+1} if $x(t)$ is a polynomial of degree less than or equal to k if the following *exactness constraints* are satisfied:

$$\sum_{i=0}^{p} a_i = 1$$

(5.50)

$$\sum_{i=1}^{p}(-i)^p a_i + j\sum_{i=-1}^{p}(-i)^{(j-1)}b_i = 1 \quad \text{for } j = 1, 2, \ldots, k$$

(5.51)

The exactness constraint of equation (5.50) is frequently referred to as the *consistency* constraint. Numerical multistep integration algorithms that satisfy equation (5.50) are said to be "consistent." For a desired polynomial of degree k, these constraints can be satisfied by a wide variety of possibilities. Several families of methods have been developed by pre-defining some of the relationships between the coefficients. The family of *Adam's* methods are defined by setting the coefficients $a_1 = a_2 = \ldots = a_p = 0$. By the consistency constraint, the coefficient a_0 must therefore equal 1.0. Thus, the Adam's methods are reduced to

$$x_{n+1} = x_n + h \sum_{i=-1}^{p} b_i f(x_{n-i}, t_{n-i}) \tag{5.52}$$

where $p = k - 1$. The Adam's methods can be further classified by the choice of implicit or explicit integration. The explicit class, frequently referred to as the "Adam's-Bashforth" methods, is specified by setting $b_{-1} = 0$ and applying the second exactness constraint as:

$$\sum_{i=0}^{k-1} (-i)^{(j-1)} b_i = \frac{1}{j} \quad j = 1, \ldots, k \tag{5.53}$$

In matrix form, equation (5.53) becomes

$$\begin{bmatrix} 1 & 1 & 1 & \cdots & 1 \\ 0 & -1 & -2 & \cdots & -(k-1) \\ 0 & 1 & 4 & \cdots & -(k-1)^2 \\ \vdots & \vdots & \vdots & \ddots & \vdots \\ 0 & (-1)^{(k-1)} & (-2)^{(k-1)} & \cdots & -(k-1)^{(k-1)} \end{bmatrix} \begin{bmatrix} b_0 \\ b_1 \\ b_2 \\ \vdots \\ b_{k-1} \end{bmatrix} = \begin{bmatrix} 1 \\ \frac{1}{2} \\ \frac{1}{3} \\ \vdots \\ \frac{1}{k-1} \end{bmatrix} \tag{5.54}$$

By choosing the desired degree k (and subsequently the order p), the remaining b_i coefficients may be found from solving equation (5.54).

Example 5.2
Find the third-order Adam's-Bashford integration method.

Solution 5.2 Setting $k = 3$ yields the following linear system:

$$\begin{bmatrix} 1 & 1 & 1 \\ 0 & -1 & -2 \\ 0 & 1 & 4 \end{bmatrix} \begin{bmatrix} b_0 \\ b_1 \\ b_2 \end{bmatrix} = \begin{bmatrix} 1 \\ \frac{1}{2} \\ \frac{1}{3} \end{bmatrix}$$

Solving this system yields

$$b_0 = \frac{23}{12}$$

$$b_1 = -\frac{16}{12}$$

$$b_2 = \frac{5}{12}$$

Thus the third-order Adam's-Bashforth method is given by:

$$x_{n+1} = x_n + \frac{1}{12}h\left[23f\left(x_n, t_n\right) - 16f\left(x_{n-1}, t_{n-1}\right) + 5f\left(x_{n-2}, t_{n-2}\right)\right] \quad (5.55)$$

When implementing this algorithm, the values of x_n, x_{n-1}, and x_{n-2} must be saved in memory. ∎

The implicit versions of the Adam's methods have $b_{-1} \neq 0$, $p = (k-2)$, are called the "Adam's-Moulton" methods, and are given by

$$x_{n+1} = x_n + h\sum_{i=-1}^{k-2} b_i f\left(x_{n-i}, t_{n-i}\right) \quad (5.56)$$

The second exactness constraint yields

$$\sum_{i=-1}^{k-2} (-i)^{(j-1)} b_i = \frac{1}{j} \quad j = 1, \ldots, k \quad (5.57)$$

or in matrix form:

$$\begin{bmatrix} 1 & 1 & 1 & 1 & \cdots & 1 \\ 1 & 0 & -1 & -2 & \cdots & -(k-2) \\ 1 & 0 & 1 & 4 & \cdots & -(k-2)^2 \\ \vdots & \vdots & \vdots & \vdots & \ddots & \vdots \\ 1 & 0 & (-1)^{(k-2)} & (-2)^{(k-2)} & \cdots & -(k-2)^{(k-2)} \end{bmatrix} \begin{bmatrix} b_{-1} \\ b_0 \\ b_1 \\ \vdots \\ b_{k-2} \end{bmatrix} = \begin{bmatrix} 1 \\ \frac{1}{2} \\ \frac{1}{3} \\ \vdots \\ \frac{1}{k} \end{bmatrix} \quad (5.58)$$

Example 5.3
Find the third-order Adam's-Moulton integration method.

Solution 5.3 Setting $k = 3$ yields the following linear system:

$$\begin{bmatrix} 1 & 1 & 1 \\ 1 & 0 & -1 \\ 1 & 0 & 1 \end{bmatrix} \begin{bmatrix} b_{-1} \\ b_0 \\ b_1 \end{bmatrix} = \begin{bmatrix} 1 \\ \frac{1}{2} \\ \frac{1}{3} \end{bmatrix}$$

Solving this system yields

$$b_{-1} = \frac{5}{12}$$

$$b_0 = \frac{8}{12}$$

$$b_1 = -\frac{1}{12}$$

Thus the third-order Adam's-Moulton method is given by:

$$x_{n+1} = x_n + \frac{1}{12}h\left[5f\left(x_{n+1}, t_{n+1}\right) + 8f\left(x_n, t_n\right) - f\left(x_{n-1}, t_{n-1}\right)\right] \quad (5.59)$$

When implementing this algorithm, the values of x_n and x_{n-1} must be saved in memory and the equations must be solved iteratively if the function $f(x)$ is nonlinear. ■

The Adam's-Moulton method is implicit and must be solved iteratively using the Newton-Raphson method (or other similar method) as shown in equation (5.48). Iterative methods require an initial value for the iterative process to reduce the number of required iterations. The explicit Adam's-Bashforth method is frequently used to estimate the initial value for the implicit Adam's-Moulton method. If sufficiently high-order predictor methods are used, the Adam's-Moulton method iteration will typically converge in only one iteration. This process is often called a *predictor-corrector* approach; the Adam's-Bashforth method predicts the solution and the implicit Adam's-Moulton corrects the solution.

Another implementation issue for multistep methods is how to start up the integration at the beginning of the simulation since a high order method requires several previous values. The usual procedure is to use a high-order one-step method or to increase the number of steps of the method with each time step to generate the required number of values for the desired multistep method.

5.2.2 Gear's methods

Another well-known family of multistep methods are the Gear's methods [9]. This family of methods is particularly well suited for the numerical solution of stiff systems. As opposed to the Adam's family of methods where all the a_i coefficients except a_0 are zero, Gear's methods are identified by having all of the b_i coefficients equal to zero except b_{-1}. Obviously since $b_{-1} \neq 0$, all Gear's methods are implicit methods. The k-th order Gear's algorithm is obtained by setting $p = k - 1$ and $b_0 = b_1 = \ldots = 0$ yielding

$$x_{n+1} = a_0 x_n + a_1 x_{n-1} + \ldots + a_{k-1} x_{n-k+1} + h b_{-1} f\left(x_{n+1}, t_{n+1}\right) \quad (5.60)$$

The $k + 1$ coefficients can be calculated explicitly by applying the exactness constraints as illustrated with the Adam's methods:

$$
\begin{bmatrix}
1 & 1 & 1 \ldots & 1 & 0 \\
0 & -1 & -2 \ldots & -(k-1) & 1 \\
0 & 1 & 4 \ldots & [-(k-1)]^2 & 2 \\
\vdots & \vdots & \vdots \ddots & \vdots & \vdots \\
0 & (-1)^k & (-2)^k \ldots & [-(k-1)]^k & k
\end{bmatrix}
\begin{bmatrix}
a_0 \\
a_1 \\
a_2 \\
\vdots \\
b_{-1}
\end{bmatrix}
=
\begin{bmatrix}
1 \\
1 \\
1 \\
\vdots \\
1
\end{bmatrix}
\quad (5.61)
$$

The solution of equation (5.61) uniquely determines the $k + 1$ coefficients of the k-th order Gear's method.

Example 5.4
Find the third-order Gear's integration method.

Solution 5.4 Setting $k = 3$ yields the following linear system:

$$\begin{bmatrix} 1 & 1 & 1 & 0 \\ 0 & -1 & -2 & 1 \\ 0 & 1 & 4 & 2 \\ 0 & -1 & -8 & 3 \end{bmatrix} \begin{bmatrix} a_0 \\ a_1 \\ a_2 \\ b_{-1} \end{bmatrix} = \begin{bmatrix} 1 \\ 1 \\ 1 \\ 1 \end{bmatrix}$$

Solving this system yields

$$b_{-1} = \frac{6}{11}$$
$$a_0 = \frac{18}{11}$$
$$a_1 = -\frac{9}{11}$$
$$a_2 = \frac{2}{11}$$

Thus the third-order Gear's method is given by:

$$x_{n+1} = \frac{18}{11}x_n - \frac{9}{11}x_{n-1} + \frac{2}{11}x_{n-2} + \frac{6}{11}hf\left(x_{n+1}, t_{n+1}\right) \tag{5.62}$$

When implementing this algorithm, the values of x_n through x_{n-2} must be saved in memory and the equations must be solved iteratively if the function $f(x)$ is nonlinear. ■

5.3   Accuracy and Error Analysis

The accuracy of numerical integration methods is impacted by two primary causes: computer round-off error and truncation error. Computer round-off error occurs as a result of the finite precision of the computer upon which the algorithm is implemented and little can be done to reduce this error short of using a computer with greater precision. A double precision word length is normally used for scientific computation. The difference between the exact solution and the calculated solution is dominated by truncation error, which arises from the truncation of the Taylor series or polynomial being used to approximate the solution.

In the implementation of numerical integration algorithms, the most effective methods are those methods that require the least amount of calculation to yield the most accurate results. In general, higher order methods produce the most accurate results, but also require the greatest amount of computation. Therefore, it is desirable to compute the solution as infrequently as possible by taking the largest time step possible between intervals. Several factors impact the size of the time step including the error introduced at each step by the numerical integration process itself. This error is the *local truncation error* (LTE) and arises from the truncation of the polynomial approximation and/or the truncation of the Taylor series expansion depending on the method used. The term *local* emphasizes that the error is introduced locally and is not residual global error from earlier time steps. The error introduced at a single step of an integration method is given by

$$\varepsilon_T \overset{\Delta}{=} x\left(t_{n+1}\right) - x_{n+1} \tag{5.63}$$

where $x\left(t_{n+1}\right)$ is the exact solution at time t_{n+1} and x_{n+1} is the numerical approximation. This definition assumes that this is the error *introduced in one step*, therefore $x(t_n) = x_n$. The local truncation error is shown graphically in Figure 5.4. To compute the error, the solution $x\left(t_{n-i}\right)$ is expanded about t_{n+1}:

$$x_{n-i} = x\left(t_{n-i}\right) = \sum_{j=0}^{\infty} \frac{\left(t_{n-i} - t_{n+1}\right)^j}{j!} \frac{d^{(j)}}{dt} x\left(t_{n+1}\right) \tag{5.64}$$

Recall that

$$f\left(x_{n-i}, t_{n-i}\right) = \dot{x}\left(t_{n-i}\right)$$
$$= \sum_{j=0}^{\infty} \frac{\left(t_{n-i} - t_{n+1}\right)^j}{j!} \frac{d^{(j+1)}}{dt} x\left(t_{n+1}\right) \tag{5.65}$$

Solving for $x\left(t_{n+1}\right) - x_{n+1}$ yields

$$\varepsilon_T = C_0 x\left(t_n\right) + C_1 x\left(t_{n-1}\right)$$
$$+ C_2 x\left(t_{n-2}\right) + \ldots + C_k x\left(t_{n-k}\right) + C_{k+1} x\left(t_{n-k-1}\right) + \ldots \tag{5.66}$$

If the order of this method is k, then the first k coefficients are equal to zero and the local truncation error is given by

$$\varepsilon_T = C_{k+1} h^{k+1} x^{(k+1)}\left(t_{n+1}\right) + O\left(h^{k+2}\right) \tag{5.67}$$

where $O\left(h^{k+2}\right)$ indicates an error on the order of h^{k+2}.

Example 5.5
Find expressions for the local truncation error of the forward-Euler, backward-Euler, and trapezoidal integration methods.

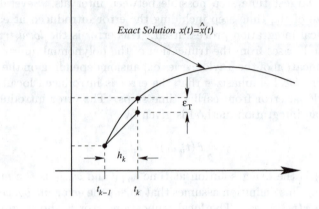

FIGURE 5.4
Graphical depiction of local truncation error

Solution 5.5

Forward Euler
Recall that the expression for the forward-Euler integration algorithm is

$$x_{n+1} = x_n + hf(x_n, t_n)$$

If $x_n = x(t_n)$ (by the definition of the local truncation error), then

$$x_n = x(t_{n+1}) - h\dot{x}(t_{n+1}) + \frac{1}{2!}h^2\ddot{x}(t_{n+1}) + \dots \qquad (5.68)$$

and

$$f(x_n, t_n) = \dot{x}(t_n) = \dot{x}(t_{n+1}) - h\ddot{x}(t_{n+1}) + \dots \qquad (5.69)$$

Thus

$$\varepsilon_T = x(t_{n+1}) - x_{n+1} \qquad (5.70)$$

$$= x(t_{n+1}) - x_n - hf(x_n, t_n) \qquad (5.71)$$

$$= x(t_{n+1}) - \left[x(t_{n+1}) - h\dot{x}(t_{n+1}) + \frac{1}{2!}h^2\ddot{x}(t_{n+1}) + \dots\right]$$

$$-h\left[\dot{x}(t_{n+1}) - h\ddot{x}(t_{n+1}) + \dots\right] \qquad (5.72)$$

$$= \frac{h^2}{2}\ddot{x}(t_{n+1}) + O(h^3) \qquad (5.73)$$

Backward Euler

The expression for the backward-Euler integration algorithm is

$$x_{n+1} = x_n + hf(x_{n+1}, t_{n+1})$$

Following the same approach as the forward-Euler method, but using

$$f(x_{n+1}, t_{n+1}) = \dot{x}(t_{n+1}) \tag{5.74}$$

then

$$\varepsilon_T = x(t_{n+1}) - x_{n+1} \tag{5.75}$$

$$= x(t_{n+1}) - x_n - hf(x_{n+1}, t_{n+1}) \tag{5.76}$$

$$= x(t_{n+1}) - \left[x(t_{n+1}) - h\dot{x}(t_{n+1}) + \frac{1}{2!}h^2\ddot{x}(t_{n+1}) + \ldots \right]$$

$$-h\dot{x}(t_{n+1}) \tag{5.77}$$

$$= -\frac{h^2}{2}\ddot{x}(t_{n+1}) - O(h^3) \tag{5.78}$$

Note that the local truncation errors for the forward- and backward-Euler methods are equal, but opposite in sign. This property is consistent with the results of Example 5.1 shown in Figure 5.2, where the respective errors were identical except in sign.

Trapezoidal

The expression for the second order trapezoidal integration algorithm is

$$x_{n+1} = x_n + \frac{1}{2}h\left[f(x_{n+1}, t_{n+1}) + f(x_n, t_n)\right]$$

Following the same approach as the previous methods using similar substitutions, then

$$\varepsilon_T = x(t_{n+1}) - x_{n+1} \tag{5.79}$$

$$= x(t_{n+1}) - x_n - \frac{1}{2}hf(x_n, t_n) - \frac{1}{2}hf(x_{n+1}, t_{n+1}) \tag{5.80}$$

$$= x(t_{n+1}) - \left[x(t_{n+1}) - h\dot{x}(t_{n+1}) + \frac{h^2}{2}\ddot{x}(t_{n+1}) - \frac{h^3}{3!}x^{(3)}(t_{n+1}) + \ldots \right]$$

$$-\frac{h}{2}\left[\dot{x}(t_{n+1}) - h\ddot{x}(t_{n+1}) + \frac{h^2}{2}x^{(3)}(t_{n+1}) + \ldots \right] - \frac{h}{2}\dot{x}(t_{n+1}) \tag{5.81}$$

$$= \frac{h^3}{6}x^{(3)}(t_{n+1}) - \frac{h^3}{4}x^{(3)}(t_{n+1}) + O(h^4) \tag{5.82}$$

$$= -\frac{1}{12}h^3 x^{(3)}(t_{n+1}) + O(h^4) \quad \blacksquare \tag{5.83}$$

Both the first order Euler methods had errors on the order of h^2, whereas the second order method (trapezoidal) had an error on the order of h^3. Both methods are implicit and must be solved iteratively at each time step. Consider the iterative solution of the trapezoidal method repeated here from equation (5.48):

$$x_{n+1}^{k+1} = x_{n+1}^k - \left[I - \frac{h}{2}\frac{\partial f}{\partial x}\right]^{-1}\bigg|_{x_{n+1}^k} \left(x_{n+1}^k - x_n - \frac{h}{2}\left[f\left(x_n\right) + f\left(x_{n+1}^k\right)\right]\right)$$

$$(5.84)$$

Similarly, the iterative solution of the backward-Euler method is given by

$$x_{n+1}^{k+1} = x_{n+1}^k - \left[I - h\frac{\partial f}{\partial x}\right]^{-1}\bigg|_{x_{n+1}^k} \left(x_{n+1}^k - x_n - h\left[f\left(x_{n+1}^k\right)\right]\right) \qquad (5.85)$$

Note that both methods require the same function evaluations and comparable computational effort, yet the trapezoidal method yields a much smaller local truncation error for the same time step size h. For this reason, the trapezoidal method is a more widely used general purpose implicit numerical integration algorithm than the backward-Euler method.

For multistep methods, a generalized expression for the local truncation error has been developed [3]. For a multistep method

$$x_{n+1} = \sum_{i=0}^{p} a_i x_{n-i} + h \sum_{i=-1}^{p} b_i f\left(x_{n-i}, t_{n-i}\right) \qquad (5.86)$$

which is exact for a polynomial solution of degree less than or equal to k, the local truncation error is given by:

$$\varepsilon_T = C_k x^{(k)}(\tau) h^{k+1} = O\left(h^{k+1}\right) \qquad (5.87)$$

where $-ph \leq \tau \leq h$ and

$$C_k \triangleq \frac{1}{(k+1)!}\left\{(p+1)^{k+1} - \left[\sum_{i=0}^{p-1} a_i\left(p-i\right)^{k+1} + (k+1)\sum_{i=-1}^{p-1} b_i\left(p-i\right)^k\right]\right\}$$

$$(5.88)$$

This expression provides an approach for approximating the local truncation error at each time step as a function of h and x.

5.4 Numerical Stability Analysis

From the previous discussion, it was shown that the choice of integration step size directly impacts the numerical accuracy of the solution. Less obvious is how the choice of step size impacts the numerical stability of integration method. Numerical stability guarantees that the global truncation error remains bounded. This guarantees that the error introduced at each time step does not accrue with time, but rather dissipates such that the choice of step size can be made by considering the local truncation error only. To analyze the effect of step size on the numerical stability of integration methods, consider the simple, scalar ODE:

$$\dot{x} = f(x) = \lambda x(t) \quad x_0 = x(t_0) \tag{5.89}$$

By inspection, the solution to this equation is

$$x(t) = x_0 e^{(\lambda t)} \tag{5.90}$$

If $\lambda < 0$, then $x(t)$ approaches zero as t goes to infinity. Conversely, if $\lambda > 0$, then $x(t)$ approaches infinity as t goes to infinity. Numerical stability ensures that the global behavior of the estimated system matches that of the actual system. Consider the forward-Euler method applied to the scalar system of equation (5.89):

$$\begin{aligned} x_{n+1} &= x_n + h\lambda x_n \\ &= (1 + h\lambda)x_n \end{aligned}$$

thus

$$\begin{aligned} x_1 &= (1 + h\lambda)x_0 \\ x_2 &= (1 + h\lambda)x_1 = (1 + h\lambda)^2 x_0 \\ &\vdots \\ x_n &= (1 + h\lambda)^n x_0 \end{aligned}$$

If $\lambda < 0$, then $x(t)$ should approach zero as t goes to infinity. This will be achieved only if

$$|1 + h\lambda| < 1 \tag{5.91}$$

Therefore, this system is stable for $\lambda < 0$ only if $h\lambda$ lies within the unit circle centered at (-1, 0) shown in Figure 5.5. Thus the larger the value of λ, the smaller the integration step size must be.

Similarly, consider the backward-Euler integration method applied to the same scalar ODE system:

$$\begin{aligned} x_{n+1} &= x_n + h\lambda x_{n+1} \\ &= \frac{x_n}{(1 - h\lambda)} \end{aligned}$$

thus

$$x_1 = \frac{x_0}{(1 - h\lambda)}$$

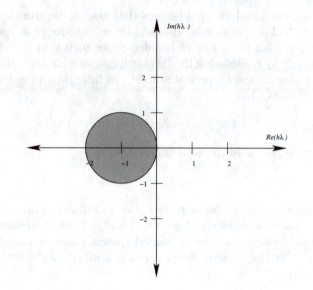

FIGURE 5.5

Region of absolute stability of the forward-Euler method

$$x_2 = \frac{x_1}{(1 - h\lambda)} = \frac{x_0}{(1 - h\lambda)^2}$$

$$\vdots$$

$$x_n = \frac{x_0}{(1 - h\lambda)^n}$$

If $\lambda < 0$, then $x(t)$ should approach zero as t goes to infinity. This will be achieved only if

$$|1 - h\lambda| > 1 \tag{5.92}$$

Therefore, this system is stable for $\lambda < 0$ only if $h\lambda$ does not lie within the unit circle centered at $(1, 0)$ shown in Figure 5.6. This implies that for all $\lambda < 0$, the backward-Euler method is numerically stable. Thus if the ODE system is stable, the integration step size may be chosen arbitrarily large without affecting the numerical stability of the solution. Thus the selection of integration step size will be dependent only on the local truncation error. Note that if $h\lambda$ is large, then x_n will rapidly approach zero. This characteristic manifests itself as a tendency to over-damp the numerical solution. This characteristic was illustrated in Figure 5.1.

Extending this approach to the general family of multistep methods yields

$$x_{n+1} = \sum_{i=0}^{p} a_i x_{n-i} + h\lambda \sum_{i=-1}^{p} b_i x_{n-i} \tag{5.93}$$

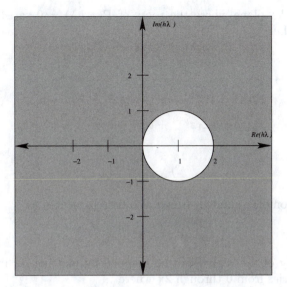

FIGURE 5.6
Region of absolute stability of the backward-Euler method

Rearranging the terms of the multistep method gives

$$x_{n+1} = \frac{(a_0 + h\lambda b_0)}{(1 - h\lambda b_{-1})} x_n + \frac{(a_1 + h\lambda b_1)}{(1 - h\lambda b_{-1})} x_{n-1} + \ldots$$

$$+ \frac{(a_p + h\lambda b_p)}{(1 - h\lambda b_{-1})} x_{n-p} \tag{5.94}$$

$$= \gamma_0 x_n + \gamma_1 x_{n-1} + \ldots + \gamma_p x_{n-p} \tag{5.95}$$

This relationship specifies the characteristic equation

$$P(z, h\lambda) = z^{p+1} + \gamma_0 z^p + \ldots + \gamma_p = 0 \tag{5.96}$$

where $z_1, z_2, \ldots, z_{p+1}$ are the (complex) roots of equation (5.96). Therefore

$$x_{n+1} = \sum_{i=1}^{p+1} C_i z_i^{n+1} \tag{5.97}$$

If $\lambda < 0$, the solution x_{n+1} will go to zero as n goes to infinity only if $|z_j| < 1$ for all $j = 1, 2, \ldots, p+1$. Thus, a multistep method is said to be *absolutely stable* for a given value of $h\lambda$ if the roots of $P(z, h\lambda) = 0$ satisfy $|z_i| < 1$ for $i = 1, \ldots, k$. Absolute stability implies that the global error decreases with increasing n. The region of absolute stability is defined to be the region in the complex $h\lambda$ plane where the roots of $P(z, h\lambda) = 0$ satisfy $|z_i| < 1$ for $i = 1, \ldots, k$. Let

$$P(z, h\lambda) = P_a(z) - h\lambda P_b(z) = 0$$

where

$$P_a(z) \triangleq z^{p+1} - a_0 z^p - a_1 z^{p-1} - \ldots - a_p$$
$$P_b(z) \triangleq b_{-1} z^{p+1} + b_0 z^p + b_1 z^{p-1} + \ldots + b_p$$

then

$$h\lambda = \frac{P_a(z)}{P_b(z)} \tag{5.98}$$

Since z is a complex number, it can also be represented as

$$z = e^{(j\theta)}$$

The boundary of the region can be mapped by plotting $h\lambda$ in the complex plane as θ varies from 0 through 2π where

$$h\lambda(\theta) = \frac{e^{j(p+1)\theta} - a_0 e^{jp\theta} - a_1 e^{j(p-1)\theta} - \ldots - a_{p-1} e^{j\theta} - a_p}{b_{-1} e^{j(p+1)\theta} + b_0 e^{jp\theta} + b_1 e^{j(p-1)\theta} + \ldots + b_{p-1} e^{j\theta} + b_p} \tag{5.99}$$

Example 5.6
Plot the regions of absolute stability of the Gear's third order and the Adam's third order (both implicit and explicit) methods.

Solution 5.6

Gear's

The region of stability of a Gear's method can be found by plotting $h\lambda$ in the complex plane from equation (5.99) by setting $p = k-1$ and $b_0, b_1, \ldots = 0$.

$$h\lambda(\theta) = \frac{e^{jk\theta} - a_0 e^{j(k-1)\theta} - \ldots - a_{k-1}}{b_{-1} e^{jk\theta}} \tag{5.100}$$

Substituting in the coefficients for the third order Gear's method yields the following expression in θ:

$$h\lambda(\theta) = \frac{e^{j3\theta} - \frac{18}{11} e^{j2\theta} + \frac{9}{11} e^{j\theta} - \frac{2}{11}}{\frac{6}{11} e^{j3\theta}} \tag{5.101}$$

By varying θ from zero to 2π, the region of absolute stability of the Gear's third-order method is shown as the shaded region of Figure 5.7.

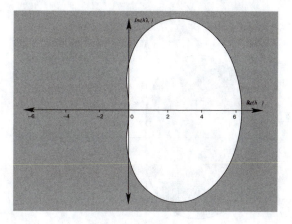

FIGURE 5.7
Region of absolute stability for Gear's third-order method

Adam's-Moulton

The region of absolute stability of the Adam's-Moulton methods can be developed from equation (5.99) by setting $p = k - 1$, and $a_1, a_2, \ldots = 0$:

$$h\lambda(\theta) = \frac{e^{jk\theta} - a_0 e^{j(k-1)\theta}}{b_{-1}e^{jk\theta} + b_0 e^{j(k-1)\theta} + b_1 e^{j(k-2)\theta} + \ldots + b_{k-2}e^{j\theta}} \qquad (5.102)$$

After substituting in the third-order coefficients, the expression for the region of absolute stability as a function of θ is given by

$$h\lambda(\theta) = \frac{e^{j3\theta} - e^{j2\theta}}{\frac{5}{12}e^{j3\theta} + \frac{8}{12}e^{j2\theta} - \frac{1}{12}e^{j\theta}} \qquad (5.103)$$

The region of absolute stability of the Adam's-Moulton third-order method is shown as the shaded region of Figure 5.8.

Adam's-Bashforth

The stability of the family of Adam's-Bashforth methods can be derived from equation (5.99) by setting $p = k - 1$, $b_{-1} = 0$, and $a_1, a_2, \ldots, = 0$:

$$h\lambda(\theta) = \frac{e^{jk\theta} - a_0 e^{j(k-1)\theta}}{b_0 e^{j(k-1)\theta} + b_1 e^{j(k-2)\theta} + \ldots + b_{k-1}} \qquad (5.104)$$

$$h\lambda(\theta) = \frac{e^{j3\theta} - e^{j2\theta}}{\frac{23}{12}e^{j2\theta} - \frac{16}{12}e^{j\theta} + \frac{5}{12}} \qquad (5.105)$$

The region of absolute stability of the Adam's-Bashforth method is shown as the shaded region in Figure 5.9.

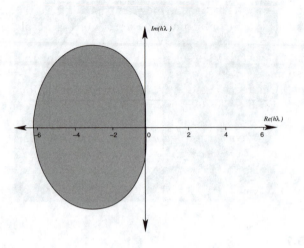

FIGURE 5.8
Region of absolute stability for Adam's-Moulton third-order method

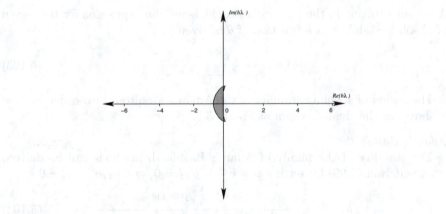

FIGURE 5.9
Region of absolute stability for Adam's-Bashforth third-order method

This example illustrates one of the primary differences between implicit and explicit methods. For the same order, the two implicit methods (Gear's and Adam's-Moulton) exhibit much larger regions of absolute stability than does the explicit method (Adam's-Bashforth). The Gear's region of absolute stability contains nearly the entire left half $h\lambda$ plane; thus, for any stable dynamic system, the step size can be chosen as large as desired without any consideration for numerical stability. Even the Adam's-Moulton region of absolute stability is quite large compared to the Adam's-Bashforth. Typically, the region of stability of explicit methods is much smaller than the region of absolute stability of corresponding order implicit methods. For this reason, implicit methods are frequently used in commercial integration packages so that the integration step size can be chosen based purely on the local truncation error criteria. The region of absolute stability shrinks as the order of the method increases, whereas the accuracy increases. This is a trade-off between accuracy, stability, and numerical efficiency in integration algorithms. ∎

5.5   Stiff Systems

Gear's methods were originally developed for the solution of systems of *stiff* ordinary differential equations. Stiff systems are systems that exhibit a wide-range of time varying dynamics from "very fast" to "very slow." A stiff linear system exhibits eigenvalues that span several orders of magnitude. A nonlinear system is stiff if its associated Jacobian matrix exhibits widely separated eigenvalues when evaluated at the operating points of interest. For efficient and accurate solutions of stiff differential equations, it is desirable for a multistep method to be "stiffly stable." An appropriate integration algorithm will allow the step size to be varied over a wide range of values and yet will remain numerically stable. A stiffly stable method exhibits the three stability regions shown in Figure 5.10 such that:

1. Region I is a region of absolute stability

2. Region II is a region of accuracy and stability

3. Region III is a region of accuracy and relative stability

Only during the initial period of the solution of the ODE do the large negative eigenvalues significantly impact the solution, yet they must be accounted for throughout the whole solution. Large negative eigenvalues ($\lambda < 0$) will decay rapidly by a factor of $1/e$ in time $1/\lambda$. If $h\lambda = \gamma + j\beta$, then the change in magnitude in one step is e^γ. If $\gamma \le \delta \le 0$, where δ defines the interface between Regions I and II, then the component is reduced by at least e^δ in one

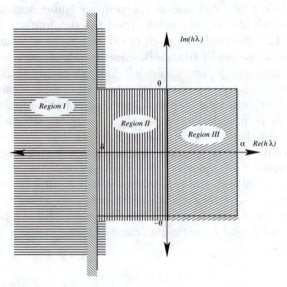

FIGURE 5.10
Regions required for stiff stability

step. After a finite number of steps, the impact of the fast components is negligible and their numerical accuracy is unimportant. Therefore the integration method is required to be absolutely stable in Region I.

Around the origin, numerical accuracy becomes more significant and relative or absolute stability is required. A region of *relative stability* consists of those values of $h\lambda$ for which the extraneous eigenvalues of the characteristic polynomial of equation (5.96) are less in magnitude than the principal eigenvalue. The principal eigenvalue is the eigenvalue which governs the system response most closely. If the method is relatively stable in Region III, then the system response will be dominated by the principal eigenvalue in that Region. If $\gamma > \alpha > 0$, one component of the system response is increasing by at least e^α one step. This increase must be limited by choosing the step sizes small enough to track this change.

If $\|\beta\| > \theta$, there are at least $\theta/2\pi$ complete cycles of oscillation in one step. Except in Region I where the response is rapidly decaying, and where $\gamma > \alpha$ is not used, the oscillatory responses must be captured. In practice, it is customary to have eight or more time points per cycle (to accurately capture the magnitude and frequency of the oscillation); thus, θ is chosen to be bounded by $\pi/4$ in Region II.

Examination of the family of Adam's-Bashforth methods shows that they all fail to satisfy the criteria to be stiffly stable and are not suitable for integrating stiff systems. Only the first and second order Adam's-Moulton (backward-Euler and trapezoidal, respectively) satisfy the stiffly stable criteria. Gear's algorithms on the other hand were developed specifically to address stiff system integration [9]. Gear's algorithms up to order six satisfy the stiff properties with the following choice of δ [3]:

Order	δ
1	0
2	0
3	0.1
4	0.7
5	2.4
6	6.1

Example 5.7

Compare the application of the third-order Adam's-Bashforth, Adam's-Moulton, and Gear's method to the integration of the following system:

$$\dot{x}_1 = 48x_1 + 98x_2 \quad x_1(0) = 1 \tag{5.106}$$
$$\dot{x}_2 = -49x_1 - 99x_2 \quad x_2(0) = 0 \tag{5.107}$$

Solution 5.7 The exact solution to Example 5.7 is

$$x_1(t) = 2e^{-t} - e^{-50t} \tag{5.108}$$
$$x_2(t) = -e^{-t} + e^{-50t} \tag{5.109}$$

This solution is shown in Figure 5.11. Both states contain both fast and slow components, with the fast component dominating the initial response and the slow component dominating the longer term dynamics. Since the Gear's, Adam's-Bashforth, and Adam's-Moulton methods are multistep methods, each method is initialized using the absolutely stable trapezoidal method for the first two to three steps using a small step size.

The Adam's-Bashforth algorithm with a time step of 0.0111 seconds is shown in Figure 5.12. Note that even with a small step size of 0.0111 seconds the inherent error in the integration algorithm eventually causes the system response to exhibit numerical instabilities. The step size can be decreased to increase the stability properties, but this requires more time steps in the integration window ($t \in [0, 2]$) than is computationally necessary.

The Adam's-Moulton response to the stiff system for an integration step size of 0.15 seconds is shown in Figure 5.13. Although a much larger time step can be used for integration as compared to the Adam's-Bashforth algorithm, the Adam's-Moulton algorithm does not exhibit numerical absolute stability. For an integration step size of $h = 0.15$ seconds, the Adam's-Moulton algorithm exhibits numerical instability. Note that the solution is oscillating with growing magnitude around the exact solution.

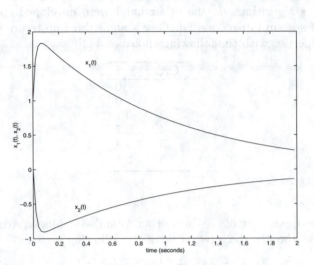

FIGURE 5.11
Stiff system response

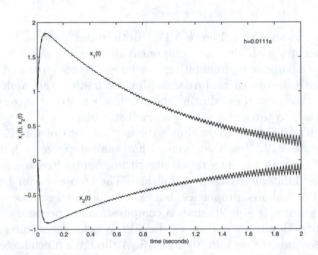

FIGURE 5.12
Adam's-Bashforth stiff system response with $h = 0.0111$ s step size

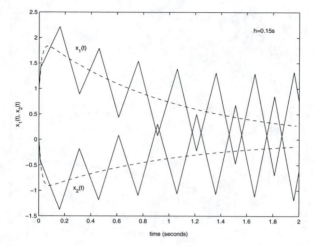

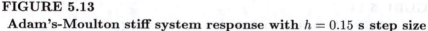

FIGURE 5.13

Adam's-Moulton stiff system response with $h = 0.15$ s step size

Figure 5.14 shows the Gear's method numerical solution to the stiff system using an integration step size of 0.15 s which is the same step size as used in the Adam's-Moulton response of Figure 5.13. The Gear's method is numerically stable since the global error decreases with increasing time.

The comparison of the three integration methods supports the necessity of using integration algorithms developed specifically for stiff systems. Even though both the third order Adam's-Bashforth and Adam's-Moulton have regions of absolute stability, these regions are not sufficient to ensure accurate stiff system integration. ∎

5.6 Step-Size Selection

For computational efficiency, it is desirable to choose the largest integration step size possible while satisfying a pre-determined level of accuracy. The level of accuracy can be maintained by constant vigilance of the local truncation error of the method if the chosen method is numerically stable. If the function $x(t)$ is varying rapidly, the step size should be chosen small enough to capture the significant dynamics. Conversely, if the function $x(t)$ is not varying significantly (nearly linear) over a finite time interval, then the integration time steps can be chosen quite large over that interval while still maintaining numerical accuracy. The true challenge to a numerical integration algorithm is when the dynamic response of $x(t)$ has intervals of both rapid variance and latency. In this case, it is desirable for the integration step size to have

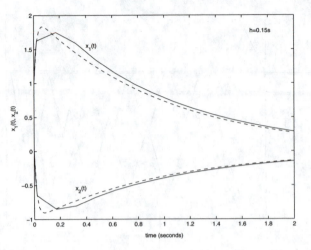

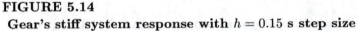

FIGURE 5.14
Gear's stiff system response with $h = 0.15$ s step size

the ability to increase or decrease throughout the simulation interval. This can be accomplished by choosing the integration step size based on the local truncation error bounds.

Consider the trapezoidal method absolute local truncation error:

$$\varepsilon_T = \frac{1}{12}h^3 x^{(3)}(\tau) \tag{5.110}$$

The local truncation error is dependent on the integration step size h and the third derivative of the function $x^{(3)}(\tau)$. If the local truncation error is chosen to be in the interval:

$$B_L \leq \varepsilon \leq B_U \tag{5.111}$$

where B_L and B_U represent the prespecified lower and upper bounds respectively, then the integration step size can be bounded by

$$h \leq \sqrt[3]{\frac{12B_U}{x^{(3)}(\tau)}} \tag{5.112}$$

If $x(t)$ is rapidly varying, the $x^{(3)}(\tau)$ will be large and h must be chosen small to satisfy $\varepsilon \leq B_U$, whereas if $x(t)$ is not varying rapidly, then $x^{(3)}(\tau)$ will be small and h can be chosen relatively large while still satisfying $\varepsilon \leq B_U$. This leads to the following procedure for calculating the integration step size:

Integration Step Size Selection

Attempt an integration step size h_{n+1} to calculate x_{n+1} from $x_n, x_{n-1}, \ldots$

1. Using x_{n+1}, calculate the local truncation error ε_T.

2. If $B_L \leq \varepsilon_T \leq B_U$, then PASS, accept h_{n+1}, $h_{next} = h_{n+1}$, and continue.

3. If $\varepsilon > B_U$, then FAIL (h_{n+1} is too large), set $h_{n+1} = \alpha h_{n+1}$, repeat integration for x_{n+1}.

4. If $\varepsilon \leq B_U$, then PASS, accept h_{n+1}, set $h_{next} = \alpha h_{n+1}$, and continue.

where

$$\alpha = \left[\frac{B_{avg}}{\varepsilon} \right]^{\frac{1}{k+1}} \tag{5.113}$$

where $B_L \leq B_{avg} \leq B_U$ and k is the degree of the method.

Some commercial integration packages implement an algorithm that is slightly different than this one. In these packages, if the local truncation error is smaller than the lower bound, then the attempted integration step size also FAILS, and the integration step is re-attempted with a larger integration step size. Once again, there is a trade off between the time spent in recalculating x_{n+1} with a larger step size and the additional computational effort acquired by simply accepting the current value and continuing on.

The difficulty in implementing this step size selection approach is the calculation of the higher order derivatives of $x(t)$. Since $x(t)$ is not known analytically, the derivatives must be calculated numerically. One common approach is to use *difference methods* to approximate the derivatives. The $(k+1)^{st}$ derivative to $x(\tau)$ is approximated by

$$x^{(k+1)}(\tau) \approx (k+1)! \nabla_{k+1} x_{n+1} \tag{5.114}$$

where $\nabla_{k+1} x_{n+1}$ is found recursively:

$$\nabla_1 x_{n+1} = \frac{x_{n+1} - x_n}{t_{n+1} - t_n} \tag{5.115}$$

$$\nabla_1 x_n = \frac{x_n - x_{n-1}}{t_n - t_{n-1}} \tag{5.116}$$

$$\vdots \tag{5.117}$$

$$\nabla_2 x_{n+1} = \frac{\nabla_1 x_{n+1} - \nabla_1 x_n}{t_{n+1} - t_{n-1}} \tag{5.118}$$

$$\nabla_2 x_n = \frac{\nabla_1 x_n - \nabla_1 x_{n-1}}{t_n - t_{n-2}} \tag{5.119}$$

$$\vdots \tag{5.120}$$

$$\nabla_{k+1} x_{n+1} = \frac{\nabla_k x_{n+1} - \nabla_k x_n}{t_{n+1} - t_{n-k}} \tag{5.121}$$

Example 5.8

Find an expression for step size selection for the trapezoidal integration method with an upper bound of 10^{-3} on the local truncation error.

Solution 5.8

The LTE for the trapezoidal method is given by

$$h \leq \sqrt[3]{\frac{12 B_U}{x^{(3)}(\tau)}} \tag{5.122}$$

The first step is to find an expression for the third derivative:

$$x^{(3)}(\tau) \approx 3! \nabla_3 x_{n+1} \tag{5.123}$$

$$\nabla_3 x_{n+1} = \frac{\nabla_2 x_{n+1} - \nabla_2 x_n}{t_{n+1} - t_{n-2}} \tag{5.124}$$

$$= \frac{\nabla_2 x_{n+1} - \nabla_2 x_n}{h_{n+1} + h_n + h_{n-1}} \tag{5.125}$$

$$= \frac{1}{h_{n+1} + h_n + h_{n-1}} \left\{ \frac{1}{h_{n+1} + h_n} \left[\frac{x_{n+1} - x_n}{h_{n+1}} - \frac{x_n - x_{n-1}}{h_n} \right] - \frac{1}{h_n + h_{n-1}} \left[\frac{x_n - x_{n-1}}{h_n} - \frac{x_{n-1} - x_{n-2}}{h_{n-1}} \right] \right\} \tag{5.126}$$

Substituting the value for B_U and the approximation for the third derivative into equation (5.122) yields the bound for h:

$$h_{n+1} \leq \frac{1}{10} \sqrt[3]{\frac{2}{\nabla_3 x_{n+1}}} \tag{5.127}$$

where $\nabla_3 x_{n+1}$ is given in equation (5.126). ■

5.7 Differential-Algebraic Systems

Many physical systems in engineering and science give rise to systems of ordinary differential equations with algebraic constraints

$$\dot{x}(t) = f(x, y, t) \tag{5.128}$$

$$0 = g(x, y, t) \tag{5.129}$$

where $x \in R^n$ are the *state* variables and $y \in R^m$ are the *algebraic* variables. These systems are sometimes referred to as descriptor systems because they arise from formulating the system equations in natural physical variables. More often, however, they are called systems of *differential-algebraic equations*

(DAEs). Often these systems are quite large and, in the past, efforts have been made to convert them to smaller, purely ordinary differential equation systems. There are several drawbacks with this approach. Firstly, the original DAE system may be sparse, but the reduced ODE system is fully connected, and therefore, not sparse. Secondly, the reduced model may eliminate physical information and there is no easy way to recover the lost information.

In many systems, it is even more difficult to find a reduced ODE system, and if one can be found, the system state variables often no longer correspond to any physical states. For these reasons, it is often desirable to leave the original system in DAE form. The solvability of a DAE system implies that there exists a unique solution for sufficiently different inputs and *consistent* initial conditions, unlike the ODE system which has a unique solution for *arbitrary* initial conditions. For a DAE system, there exists only one set of initial conditions to the non-state (algebraic) variables. The initial values that are consistent with the system input are called *admissible* initial values. The admissible initial values are those that satisfy

$$y_0 = g^{-1}(x_0, t_0) \tag{5.130}$$

The approach of applying a numerical ODE method to the DAE system was first introduced by Gear [10], and consists of replacing $\dot{x}(t)$ by a k-step backwards difference formula (BDF) approximation

$$\dot{x}(t) \approx \frac{\rho_n x_n}{h_n} = \frac{1}{h_n} \sum_{i=0}^{k} \alpha_i x_{n-i} \tag{5.131}$$

and then solving the resulting equations

$$\rho_n x_n = h_n f(x_n, y_n, t_n) \tag{5.132}$$

$$0 = g(x_n, y_n, t_n) \tag{5.133}$$

for approximations to x_n and y_n.

Various other numerical integration techniques have been studied for application to DAE systems. Variable step size/fixed formula code has been proposed to solve the system of DAEs [32]. Specifically, a classic fourth-order Runge-Kutta method was used to solve for x and a third order BDF to solve for y. In general, however, many DAE systems may be solved by any multistep numerical integration method which is convergent when applied to ODEs [15].

5.8 Power System Applications

Power systems in general are considered to be large scale, often involving several hundred equations to describe the behavior of an interconnected power system during and following a fault on the system. As power system operations become increasingly complex, it becomes necessary to be able to perform analyses of voltage conditions and system stability. A medium-sized power company serving a mixed urban and rural population of two to three million people operates a network that may typically contain hundreds of buses and thousands of transmission lines, excluding the distribution system [6]. Under certain assumptions such as instantaneous transmission lines, interconnected power systems are often modeled in DAE form with over 1000 differential equations and 10,000 algebraic constraint equations. One traditional approach to solving large scale systems of this type has been to replace the full system model with a reduced-order state space model.

5.8.1 Transient Stability Analysis

The "classical model" of a synchronous machine is often used to study the transient stability of a power system during the period of time in which the system dynamics depend largely on the stored kinetic energy in the rotating masses. This is usually on the order of a second or two. The classical model is derived under several simplifying assumptions [1]:

1. Mechanical power input, P_m, is constant.

2. Damping is negligible.

3. The constant voltage behind transient reactance model for the synchronous machines is valid.

4. The rotor angle of the machine coincides with the voltage behind the transient reactance angle.

5. Loads are represented as constant impedances.

The equations of motion are given by

$$\dot{\omega}_i = \frac{1}{M_i} \left(P_{m_i} - E_i^2 G_{ii} - E_i \sum_{j \neq i}^{n} E_j \left(B_{ij} \sin \delta_{ij} + G_{ij} \cos \delta_{ij} \right) \right) \quad (5.134)$$

$$\dot{\delta}_i = \omega_i - \omega_s \quad i = 1, \ldots, n \quad (5.135)$$

where n is the number of machines, ω_s is the synchronous angular frequency, $\delta_{ij} = \delta_i - \delta_j$, $M_i = \frac{2H_i}{\omega_s}$, and H_i is the inertia constant in seconds. B_{ij} and G_{ij} are elements of the reduced admittance matrix Y at the internal nodes of the machine. The loads are modeled as constant impedances which are then absorbed into the admittance matrix. The classical model is appropriate for frequency studies that result from faults on the transmission system for the first or second swing of the rotor angle. The procedure for setting up a transient stability analysis is given below.

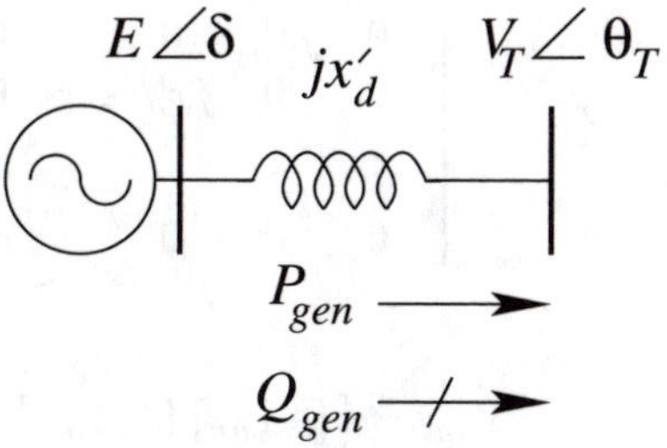

FIGURE 5.15
Voltage behind transient reactance

Transient Stability Analysis

1. Perform a load flow analysis to obtain system voltages, angles, active and reactive power generation.

2. For each generator $i, \ldots, n$, in the system, calculate the internal voltage and initial rotor angle $E\angle\delta_0$:

$$I_{gen}^* = \frac{(P_{gen} + jQ_{gen})}{V_T\angle\theta_T} \qquad (5.136)$$

$$E\angle\delta_0 = jx_d'I_{gen} + V_T\angle\theta_T \qquad (5.137)$$

where $P_{gen} + jQ_{gen}$ are the generated active and reactive power obtained from the power flow solution and $V_T\angle\theta_T$ is the generator terminal voltage.

3. For each load $1, \ldots, m$ in the system, convert the active and reactive power loads to admittances:

$$Y_L = G_L + jB_L = \frac{I_L}{V_L\angle\theta_L} \qquad (5.138)$$

$$= \frac{S_L^*}{V_L^2} \qquad (5.139)$$

$$= \frac{P_L - jQ_L}{V_L^2} \qquad (5.140)$$

Add the shunt admittance Y_L to the corresponding diagonal of the admittance matrix.

4. For each generator in the system, augment the admittance matrix by adding an internal bus connected to the terminal bus with the transient reactance x_d' as shown in Figure 5.15.

Then let

$$
Y_{nn} = \begin{bmatrix}
jx'_{d_1} & 0 & 0 & \cdots & 0 \\
0 & jx'_{d_2} & 0 & \cdots & 0 \\
0 & 0 & jx'_{d_3} & \cdots & 0 \\
\vdots & \vdots & \vdots & \ddots & \vdots \\
0 & 0 & 0 & \cdots & jx'_{d_n}
\end{bmatrix}
$$

and

$$
Y_{nm} = \left[\left[-Y_{nn} \right] \left[0_{n \times m} \right] \right]
$$
$$
Y_{mn} = Y_{nm}^T
$$

where $[0_{n \times m}]$ is an $(n \times m)$ matrix of zeros.

Further, let

$$
Y_{mm} = \left[Y_{original} + \begin{bmatrix}
Y_{nn} & 0_{n \times (m-n)} \\
0_{(m-n) \times n} & 0_{(m-n) \times (m-n)}
\end{bmatrix} \right]
$$

This matrix layout assumes that the systems buses have been ordered such that the generators are numbered $1, \ldots, n$, with the remaining load buses numbered $n + 1, \ldots, m$.

5. The reduced admittance matrix Y_{red} is found by

$$
Y_{red} = \left[Y_{nn} - Y_{nm} Y_{mm}^{-1} Y_{mn} \right] \tag{5.141}
$$
$$
= G_{red} + j B_{red} \tag{5.142}
$$

The reduced admittance matrix is now $n \times n$ where n is the number of generators, whereas the original admittance matrix was $m \times m$.

6. Repeat steps 4 and 5 to calculate the fault-on reduced admittance matrix and the post-fault reduced admittance matrix (if different from the pre-fault reduced admittance matrix).

7. For $0 < t \le t_{apply}$, integrate the transient stability equations (5.134)-(5.135) with the integration method of choice using the pre-fault reduced admittance matrix, where t_{apply} is the time the system fault is applied. In many applications $t_{apply} = 0$.

8. For $t_{apply} < t \le t_{clear}$, integrate the transient stability equations (5.134)-(5.135) with the integration method of choice using the fault-on reduced admittance matrix, where t_{clear} is the time the system fault is cleared.

9. For $t_{clear} < t \le t_{max}$, integrate the transient stability equations (5.134)-(5.135) with the integration method of choice using the post-fault reduced admittance matrix, where t_{max} is end of the simulation interval.

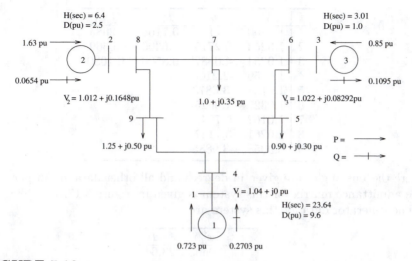

FIGURE 5.16
Three-machine, nine-bus system of Example 5.9

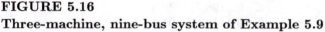

At the end of the simulation, the state variables (δ_i, ω_i) for each of the generators may be plotted against time. The rotor angle responses are in radians and may be converted to degrees if desired. The rotor angular frequency is in radians per second and may be converted to hertz (cycles per second) if desired. These waveforms may then be analyzed to determine whether or not system stability is maintained. If the system responses diverge from one another or exhibit growing oscillations, then the system is most probably unstable.

Example 5.9

For the three-machine, nine-bus system shown in Figure 5.16, a solid three-phase fault occurs at bus 8 at 0.1 seconds. The fault is cleared by opening line 8-9. Determine whether or not the system will remain stable for a clearing time of 0.12 seconds after the application of the fault.

Solution 5.9 Following the procedure for transient stability analysis outlined previously, the first step is to perform a power flow analysis of the system. The line and bus data are given in Figure 5.16. The load flow results are

i	V	θ	P_{gen}	Q_{gen}
1	1.0400	0	0.7164	0.2685
2	1.0253	9.2715	1.6300	0.0669
3	1.0254	4.6587	0.8500	-0.1080
4	1.0259	-2.2165		
5	1.0128	-3.6873		
6	1.0327	1.9625		
7	1.0162	0.7242		
8	1.0261	3.7147		
9	0.9958	-3.9885		

where the bus angles are given in degrees and all other data are in per unit. The admittance matrix for this system is given in Figure 5.17.

The generator data for this system are

i	x'_d	H
1	0.0608	23.64
2	0.1198	6.40
3	0.1813	3.01

The internal voltages and rotor angles for each generator are computed using the generated active and reactive powers, voltage magnitudes, and angles as:

$$I_1^* = \frac{(0.7164 + j0.2685)}{1.0400\angle 0°} = 0.6888 + j0.2582$$
$$E_1 \angle \delta_1 = (j0.0608)(0.6888 - j0.2582) + 1.0400\angle 0° = 1.0565\angle 2.2718°$$

$$I_2^* = \frac{(1.6300 + j0.0669)}{1.0253\angle 9.2715°} = 1.5795 - j0.1918$$
$$E_2 \angle \delta_2 = (j0.1198)(1.5795 + j0.1918) + 1.0253\angle 9.2715° = 1.0505\angle 19.7162°$$

$$I_3^* = \frac{(0.8500 - j0.1080)}{1.0254\angle 4.6587°} = 0.8177 - j0.1723$$
$$E_3 \angle \delta_3 = (j0.1813)(0.8177 + j0.1723) + 1.0254\angle 4.6587° = 1.0174\angle 13.1535°$$

The next step is to convert the loads to equivalent impedances:

$$G_5 + jB_5 = \frac{(0.90 - j0.30)}{1.0128^2} = 0.8773 - j0.2924$$
$$G_7 + jB_7 = \frac{(1.00 - j0.35)}{1.0162^2} = 0.9684 - j0.3389$$
$$G_9 + jB_9 = \frac{(1.25 - j0.50)}{0.9958^2} = 1.2605 - j0.5042$$

These values are added to the diagonal of the original admittance matrix.

$$[Y_{bus}] =$$

$$\begin{bmatrix}
17.3611\angle-90.00° & 0 & 0 & 17.3611\angle90.00° & 0 & 0 & 0 & 0 & 0 \\
0 & 16.0000\angle-90.00° & 0 & 0 & 0 & 0 & 0 & 16.0000\angle90.00° & 0 \\
0 & 0 & 17.0648\angle-90.00° & 0 & 17.0648\angle90.00° & 0 & 0 & 0 & 0 \\
17.3611\angle90.00° & 0 & 0 & 39.4478\angle-85.19° & 10.6886\angle100.47° & 0 & 0 & 0 & 11.6841\angle96.71° \\
0 & 0 & 17.0648\angle90.00° & 33.8085\angle-90.00° & 16.1657\angle-78.50° & 5.7334\angle102.92° & 0 & 0 & 0 \\
0 & 0 & 0 & 0 & 5.7334\angle102.92° & 32.2461\angle-85.67° & 9.8522\angle96.73° & 0 & 0 \\
0 & 0 & 0 & 0 & 0 & 9.8522\angle96.73° & 23.4676\angle-83.22° & 13.7931\angle96.73° & 0 \\
0 & 16.0000\angle90.00° & 0 & 0 & 0 & 0 & 13.7931\angle96.73° & 35.5564\angle-85.48° & 6.0920\angle101.24° \\
0 & 0 & 0 & 11.6841\angle96.71° & 0 & 0 & 0 & 6.0920\angle101.24° & 17.5252\angle-81.62°
\end{bmatrix}$$

FIGURE 5.17
Admittance matrix for Example 5.9

The reduced admittance matrices can now be computed as outlined in steps 4 and 5 above. The pre-fault admittance matrix is

$$Y_{red}^{pre\text{-}fault} = \begin{bmatrix} 0.8453 - j2.9881 & 0.2870 + j1.5131 & 0.2095 + j1.2257 \\ 0.2870 + j1.5131 & 0.4199 - j2.7238 & 0.2132 + j1.0880 \\ 0.2095 + j1.2257 & 0.2132 + j1.0880 & 0.2769 - j2.3681 \end{bmatrix}$$

The fault-on matrix is found similarly, except that the Y_{mm} is altered to reflect the fault on bus 8. The solid three-phase fault is modeled by shorting the bus to ground. In the admittance matrix, the row and column corresponding to bus 8 are removed. The lines between bus 8 and adjacent buses are now connected to ground; thus, they will still appear in the original admittance diagonals. The column of Y_{nm} and the row of Y_{mn} corresponding to bus 8 must also be removed. The matrix Y_{nn} remains unchanged. The fault-on reduced admittance matrix is

$$Y_{red}^{fault\text{-}on} = \begin{bmatrix} 0.6567 - j3.8159 & 0 & 0.0701 + j0.6306 \\ 0 & 0 - j5.4855 & 0 \\ 0.0701 + j0.6306 & 0 & 0.1740 - j2.7959 \end{bmatrix}$$

The post-fault reduced admittance matrix is computed in much the same way, except that line 8-9 is removed from Y_{mm}. The elements of Y_{mm} are updated to reflect the removal of the line:

$$Y_{mm}(8,8) = Y_{mm}(8,8) + Y_{mm}(8,9)$$
$$Y_{mm}(9,9) = Y_{mm}(9,9) + Y_{mm}(8,9)$$
$$Y_{mm}(8,9) = 0$$
$$Y_{mm}(9,8) = 0$$

Note that the diagonals must be updated before the off-diagonals are zeroed out. The post-fault reduced admittance is then computed:

$$Y_{red}^{post\text{-}fault} = \begin{bmatrix} 1.1811 - j2.2285 & 0.1375 + j0.7265 & 0.1909 + j1.0795 \\ 0.1375 + j0.7265 & 0.3885 - j1.9525 & 0.1987 + j1.2294 \\ 0.1909 + j1.0795 & 0.1987 + j1.2294 & 0.2727 - j2.3423 \end{bmatrix}$$

These admittance matrices are then ready to be substituted into the transient stability equations at the appropriate time in the simulation.

Applying the trapezoidal algorithm to the transient stability equations yields the following system of equations:

$$\delta_1(n+1) = \delta_1(n) + \frac{h}{2}\left[\omega_1(n+1) - \omega_s + \omega_1(n) - \omega_s\right] \qquad (5.143)$$

$$\omega_1(n+1) = \omega_1(n) + \frac{h}{2}\left[f_1(n+1) + f_1(n)\right] \qquad (5.144)$$

$$\delta_2(n+1) = \delta_2(n) + \frac{h}{2}\left[\omega_2(n+1) - \omega_s + \omega_2(n) - \omega_s\right] \qquad (5.145)$$

$$\omega_2(n+1) = \omega_2(n) + \frac{h}{2}\left[f_2(n+1) + f_2(n)\right] \tag{5.146}$$

$$\delta_3(n+1) = \delta_3(n) + \frac{h}{2}\left[\omega_3(n+1) - \omega_s + \omega_3(n) - \omega_s\right] \tag{5.147}$$

$$\omega_3(n+1) = \omega_3(n) + \frac{h}{2}\left[f_3(n+1) + f_3(n)\right] \tag{5.148}$$

where

$$f_i(n+1) = \frac{1}{M_i}\left(P_{m_i} - E_i^2 G_{ii} - E_i \sum_{j\neq i}^{n} E_j\left(B_{ij}\sin\delta_{ij}(n+1) + G_{ij}\cos\delta_{ij}(n+1)\right)\right)$$
$$\tag{5.149}$$

Since the transient stability equations are nonlinear and the trapezoidal method is an implicit method, they must be solved iteratively using the Newton-Raphson method at each time point. The iterative equations are

$$\left[I - \frac{h}{2}\left[J(n+1)^k\right]\right]\begin{bmatrix} \delta_1(n+1)^{k+1} - \delta_1(n+1)^k \\ \omega_1(n+1)^{k+1} - \omega_1(n+1)^k \\ \delta_2(n+1)^{k+1} - \delta_2(n+1)^k \\ \omega_2(n+1)^{k+1} - \omega_2(n+1)^k \\ \delta_3(n+1)^{k+1} - \delta_3(n+1)^k \\ \omega_3(n+1)^{k+1} - \omega_3(n+1)^k \end{bmatrix} =$$

$$-\left[\begin{bmatrix} \delta_1^k(n+1) \\ \omega_1^k(n+1) \\ \delta_2^k(n+1) \\ \omega_2^k(n+1) \\ \delta_3^k(n+1) \\ \omega_3^k(n+1) \end{bmatrix} - \begin{bmatrix} \delta_1(n) \\ \omega_1(n) \\ \delta_2(n) \\ \omega_2(n) \\ \delta_3(n) \\ \omega_3(n) \end{bmatrix} - \frac{h}{2}\begin{bmatrix} \omega_1^k(n+1) + \omega_1(n) - 2\omega_s \\ f_1^k(n+1) + f_1(n) \\ \omega_2^k(n+1) + \omega_2(n) - 2\omega_s \\ f_2^k(n+1) + f_2(n) \\ \omega_1^k(n+1) + \omega_1(n) - 2\omega_s \\ f_3^k(n+1) + f_3(n) \end{bmatrix}\right] \tag{5.150}$$

where

$$[J] = \begin{bmatrix} 0 & 1 & 0 & 0 & 0 & 0 \\ \frac{\partial f_1}{\partial \delta_1} & 0 & \frac{\partial f_1}{\partial \delta_2} & 0 & \frac{\partial f_1}{\partial \delta_3} & 0 \\ 0 & 0 & 0 & 1 & 0 & 0 \\ \frac{\partial f_2}{\partial \delta_1} & 0 & \frac{\partial f_2}{\partial \delta_2} & 0 & \frac{\partial f_2}{\partial \delta_3} & 0 \\ 0 & 0 & 0 & 0 & 0 & 1 \\ \frac{\partial f_3}{\partial \delta_1} & 0 & \frac{\partial f_3}{\partial \delta_2} & 0 & \frac{\partial f_3}{\partial \delta_3} & 0 \end{bmatrix} \tag{5.151}$$

Note that LU factorization must be employed to solve the discretized equations. These equations are iterated at each time point until convergence of the Newton-Raphson algorithm. The fault-on and post-fault matrices are substituted in at the appropriate times in the integration. The simulation results are shown in Figures 5.18 and 5.19 for the rotor angles and angular frequencies, respectively. From the waveforms shown in these figures, it can be concluded that the system remains stable since the waveforms do not diverge during the simulation interval. ∎

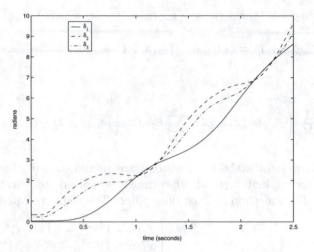

FIGURE 5.18
Rotor angle response for Example 5.9

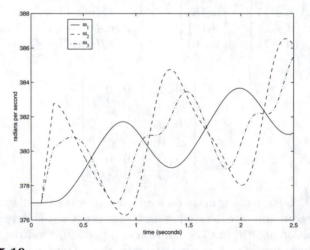

FIGURE 5.19
Angular frequency response for Example 5.9

5.9 Mid-Term Stability Analysis

Reduction processes frequently destroy the natural physical structure and sparsity of the full order system. Numerical solution algorithms which make use of structure and sparsity for efficiency perform poorly on the reduced-order system even though the reduced-order system is still quite large. After the first few seconds of a power system disturbance, the classical model representation may no longer be valid due to the dynamic behavior of the automatic voltage regulator, the turbine/governor system, under-load-tap-changing transformers, and the dynamic nature of some system loads. For mid-term stability analyses, a more detailed model is required to capture a wider range of system behavior. Since the behavior of loads may significantly impact the stability of the system, it is desirable to be able to retain individual load buses during the simulation. This type of model is often referred to as a "structure-preserving" model, since the physical structure of the power system is retained. The inclusion of the load buses requires the solution of the set of power flow equations governing the system network. This constraint leads to the inclusion of the algebraic power flow equations in conjunction with the differential equations describing the states. One example of a structure-preserving DAE model is given below:

$$T_{d0_i} \dot{E}'_{q_i} = -E'_{q_i} - \left(x_{d_i} - x'_{d_i}\right) I_{d_i} + E_{fd_i} \tag{5.152}$$

$$T_{q0_i} \dot{E}'_{d_i} = -E'_{d_i} + \left(x_{q_i} - x'_{q_i}\right) I_{q_i} \tag{5.153}$$

$$\dot{\delta}_i = \omega_i - \omega_s \tag{5.154}$$

$$\frac{2H_i}{\omega_s} \dot{\omega}_i = T_{m_i} - E'_{d_i} I_{d_i} - E'_{q_i} I_{q_i} - \left(x'_{q_i} - x'_{d_i}\right) I_{d_i} I_{q_i} \tag{5.155}$$

$$T_{E_i} \dot{E}_{fd_i} = -\left(K_{E_i} + S_{E_i}\left(E_{fd_i}\right)\right) E_{fd_i} + V_{R_i} \tag{5.156}$$

$$T_{F_i} \dot{R}_{F_i} = -R_{F_i} + \frac{K_{F_i}}{T_{F_i}} E_{fd_i} \tag{5.157}$$

$$T_{A_i} \dot{V}_{R_i} = -V_{R_i} + K_{A_i} R_{F_i} - \frac{K_{A_i} K_{F_i}}{T_{F_i}} E_{fd_i} + K_{A_i}(V_{ref_i} - V_{T_i}) \tag{5.158}$$

$$T_{RH_i} \dot{T}_{M_i} = -T_{M_i} + \left(1 - \frac{K_{HP_i} T_{RH_i}}{T_{CH_i}}\right) P_{CH_i} + \frac{K_{HP_i} T_{RH_i}}{T_{CH_i}} P_{SV_i} \tag{5.159}$$

$$T_{CH_i} \dot{P}_{CH_i} = -P_{CH_i} + P_{SV_i} \tag{5.160}$$

$$T_{SV_i} \dot{P}_{SV_i} = -P_{SV_i} + P_{C_i} - \frac{1}{R} \frac{\omega_i}{\omega_s} \tag{5.161}$$

and

$$0 = V_i e^{j\theta_i} + \left(r_s + jx'_{d_i}\right)\left(I_{d_i} + jI_{q_i}\right)e^{j\left(\delta_i - \frac{\pi}{2}\right)}$$
$$- \left[E'_{d_i} + \left(x'_{q_i} - x'_{d_i}\right)I_{q_i} + jE'_{q_i}\right]e^{j\left(\delta_i - \frac{\pi}{2}\right)} \tag{5.162}$$

$$0 = V_i e^{j\theta_i}\left(I_{d_i} - jI_{q_i}\right)e^{-j\left(\delta_i - \frac{\pi}{2}\right)} - \sum_{k=1}^{N} V_i V_k Y_{ik} e^{j(\theta_i - \theta_k - \phi_{ik})} \tag{5.163}$$

$$0 = P_i + jQ_i - \sum_{k=1}^{N} V_i V_k Y_{ik} e^{j(\theta_i - \theta_k - \phi_{ik})} \tag{5.164}$$

These equations describe the behavior of a two-axis generator model, a simple automatic voltage regulator and exciter, a simple turbine/governor, and constant power loads. This set of dynamic equations is listed here for illustration purposes and is not intended to be inclusive of all possible representations. A detailed development of these equations may be found in [30].

These equations may be modeled in a more general form as

$$\dot{x} = f(x,y) \tag{5.165}$$
$$0 = g(x,y) \tag{5.166}$$

where the state vector x contains the dynamic state variables of the generators. The vector y is typically much larger and contains all of the network variables including bus voltage magnitude and angle. It may also contain generator states such as currents that are not inherently dynamic. There are two typical approaches to solving this set of differential/algebraic equations. The first is the method suggested by Gear whereby all equations are solved simultaneously. The second approach is to solve the differential and algebraic sets of equations separately and iterate between each.

Consider the first approach applied to the DAE system using the trapezoidal integration method:

$$x(n+1) = x(n) + \frac{h}{2}\left[f\left(x(n+1), y(n+1)\right) + f\left(x(n), y(n)\right)\right] \tag{5.167}$$
$$0 = g\left(x(n+1), y(n+1)\right) \tag{5.168}$$

This set of nonlinear equations must then be solved using the Newton-Raphson method for the combined state vector $[x(n+1)\ \ y(n+1)]^T$:

$$\begin{bmatrix} I - \frac{h}{2}\frac{\partial f}{\partial x} & -\frac{h}{2}\frac{\partial f}{\partial y} \\ \frac{\partial g}{\partial x} & \frac{\partial g}{\partial y} \end{bmatrix}\begin{bmatrix} x(n+1)^{k+1} - x(n+1)^k \\ y(n+1)^{k+1} - y(n+1)^k \end{bmatrix}$$
$$= \begin{bmatrix} x(n+1)^k - x(n) - \frac{h}{2}\left[f\left(x(n+1)^k, y(n+1)^k\right) + f\left(x(n), y(n)\right)\right] \\ g\left(x(n+1)^k, y(n+1)^k\right) \end{bmatrix} \tag{5.169}$$

The advantages of this method are that since the whole set of system equations is used, the system matrices are quite sparse and sparse solution techniques can be used efficiently. Secondly, since the set of equations are solved simultaneously, the iterations are more likely to converge since the only iteration involved is that of the Newton-Raphson algorithm. Note, however, that in a system of ODEs, the left hand side matrix (the matrix to be factored in LU factorization) can be made to be diagonally dominant (and therefore well-conditioned) by decreasing the step size h. In DAE systems, however, the left hand side matrix may be ill conditioned under certain operating points where $\frac{\partial g}{\partial y}$ is ill conditioned causing difficulty in solving the system of equations. This situation may occur during the simulation of voltage collapse where the states encounter a bifurcation. The subject of bifurcation and voltage collapse is complex and outside the scope of this book; however, several excellent texts have been published that study these phenomena in great detail [17] [30] [42].

The second approach to solving the system of DAEs is to solve each of the subsystems independently and iteratively. The system of differential equations is solved first for $x(n + 1)$ while holding $y(n + 1)$ constant as an input. After $x(n + 1)$ has been found, it is then used as an input to solve the algebraic system. The updated value of $y(n + 1)$ is then substituted back into the differential equations, and $x(n+1)$ is recalculated. This back and forth process is repeated until the values $x(n+1)$ and $y(n+1)$ converge. The solution is then advanced to the next time point. The advantage of this method is simplicity in programming, since each subsystem is solved independently and the Jacobian elements $\frac{\partial f}{\partial y}$ and $\frac{\partial g}{\partial x}$ are not used. In some cases, this may speed up the computation, although more iterations may be required to reach convergence.

5.10   Problems

1. Determine a Taylor-series expansion for the solution of the equation

$$\dot{x} = x^2 \quad x(0) = 1$$

about the point $\hat{x} = 0$ (a McClaurin series expansion). Use this approximation to compute x for $\hat{x} = 0.2$ and $\hat{x} = 1.2$. Compare with the exact solution and explain the results.

2. Use the following algorithms to solve the initial value problem

$$\dot{x}_1 = -2x_2 + 2t^2 \quad x_1(0) = -2$$
$$\dot{x}_2 = \frac{1}{2}x_1 + 2t \quad x_2(0) = 0$$

   (a) Backward Euler
   (b) Forward Euler
   (c) Trapezoidal Rule
   (d) Fourth Order Runge Kutta

   Compare the answers to the exact solution

$$x_1(t) = -2\cos t$$
$$x_2(t) = -2\sin t + t^2$$

3. Consider a simple ecosystem consisting of rabbits that have an infinite food supply and foxes that prey upon the rabbits for their food. A classical mathematical "predator-prey" model due to Volterra describes this system by a pair of non-linear, first-order differential equations:

$$\dot{r} = \alpha r + \beta r f \quad r(0) = r_0$$
$$\dot{f} = \gamma f + \delta r f \quad f(0) = f_0$$

   where $r = r(t)$ is the number of rabbits, $f = f(t)$ is the number of foxes. When $\beta = 0$, the two populations do not interact, and so the rabbits multiply and the foxes die off from starvation.

   Investigate the behavior of this system for $\alpha = -1, \beta = 0.01, \gamma = 0.25$, and $\delta = -0.01$. Use the trapezoidal integration method with $h = 0.1$ and $T = 50$, and the set of initial conditions $r_0 = 30 \pm 10$ and $f_0 = 80 \pm 10$. Plot (1) t vs. r and f, and (2) r vs. f for each case.

4. Consider the following linear multistep formula

$$x_{n+1} = a_0 x_n + a_1 x_{n-1} + a_2 x_{n-2} + a_3 x_{n-3} + h b_{-1} f(x_{n+1}, t_{n+1})$$

   (a) What are the number of steps in the formula?

   (b) What is the maximum order that will enable you to determine all the coefficients of the formula?

   (c) Assuming uniform step size h, find all the coefficients of the formula such that its order is the answer of part (b).

   (d) Is the formula implicit or explicit?

   (e) Assuming uniform step size h, find the expression for the Local Truncation error of the formula.

   (f) Is the formula absolutely stable?

   (g) Is the formula stiffly-stable?

5. Consider the following initial value problem:

$$\dot{x} = 100(\sin(t) - x), \quad x(0) = 0;$$

   The exact solution is:

$$x(t) = \frac{\sin(t) - 0.01\cos(t) + 0.01e^{-100t}}{1.0001}$$

   Solve the initial value problem with $h = 0.02$ s using the following integration methods. Plot each of the numerical solutions against the exact solution over the range $t \in [0, 3.0]$ seconds. Plot the global error for each method over the range $t \in [0, 3.0]$ seconds. Discuss your results.

   (a) Backward Euler

   (b) Forward Euler

   (c) Trapezoidal Rule

   (d) Fourth Order Runge Kutta

   (e) Repeat (a)-(d) with step size $h = 0.03$ s.

6. Consider the following system of "stiff" equations:

$$\dot{x}_1 = -2x_1 + x_2 + 100 \quad x_1(0) = 0$$
$$\dot{x}_2 = 10^4 x_1 - 10^4 x_2 + 50 \quad x_2(0) = 0$$

(a) Determine the *maximum step size* h_{max} for the forward Euler algorithm to remain numerically stable.

(b) Use the forward Euler algorithm with step size $h = \frac{1}{2} h_{max}$ to solve for $x_1(t)$ and $x_2(t)$ for $t > 0$.

(c) Repeat using a step size $h = 2h_{max}$.

(d) Use the backward Euler algorithm to solve the stiff equations. Choose the following step sizes in terms of the maximum step size h_{max}.

   i. $h = 10h_{max}$
   ii. $h = 100h_{max}$
   iii. $h = 1000h_{max}$
   iv. $h = 10,000h_{max}$

(e) Repeat using the multistep method (Gear's method) of Problem 4.

7. Consider a multistep method of the form

$$x_{n+2} - x_{n-2} + \alpha(x_{n+1} - x_{n-1}) = h\left[\beta(f_{n+1} + f_{n-1}) + \gamma f_n\right]$$

(a) Show that the parameters α, β, and γ can be chosen uniquely so that the method has order $p = 6$.

(b) Discuss the stability properties of this method. For what region is it absolutely stable? Stiffly stable?

8. A power system can be described by the following ODE system:

$$\dot{\delta}_i = \omega_i - \omega_s$$
$$M_i \dot{\omega}_i = P_i - E_i \sum_{k=1}^{n} E_k Y_{ik} \sin(\delta_i - \delta_k - \phi_{ik})$$

Using a step size of $h = 0.01$ s and the trapezoidal integration method, determine whether or not this system is stable for a clearing time of 4 cycles.

Let

$$E_1 = 1.0566\angle 2.2717°$$
$$E_2 = 1.0502\angle 19.7315°$$
$$E_3 = 1.0170\angle 13.1752°$$
$$P_1 = 0.716$$
$$P_2 = 1.630$$
$$P_3 = 0.850$$

and the following admittance matrices:

Prefault:
$$0.846 - j2.988 \quad 0.287 + j1.513 \quad 0.210 + j1.226$$
$$0.287 + j1.513 \quad 0.420 - j2.724 \quad 0.213 + j1.088$$
$$0.210 + j1.226 \quad 0.213 + j1.088 \quad 0.277 - j2.368$$

Fault-on:
$$0.657 - j3.816 \quad 0.000 + j0.000 \quad 0.070 + j0.631$$
$$0.000 + j0.000 \quad 0.000 - j5.486 \quad 0.000 + j0.000$$
$$0.070 + j0.631 \quad 0.000 + j0.000 \quad 0.174 - j2.796$$

Post-Fault:
$$1.181 - j2.229 \quad 0.138 + j0.726 \quad 0.191 + j1.079$$
$$0.138 + j0.726 \quad 0.389 - j1.953 \quad 0.199 + j1.229$$
$$0.191 + j1.079 \quad 0.199 + j1.229 \quad 0.273 - j2.342$$

9. A double pendulum system is shown in Figure 5.20. Masses m_1 and m_2 are connected by massless rods of length r_1 and r_2. The equations of motion of the two masses, expressed in terms of the angles θ_1 and θ_2 as indicated, are:

$$-(m_1 + m_2)gr_1 \sin\theta_1 = (m_1 + m_2)r_1^2\ddot{\theta}_1 + m_2r_1r_2\ddot{\theta}_2 \cos(\theta_1 - \theta_2)$$
$$+ m_2r_1r_2\dot{\theta}_2^2 \sin(\theta_1 - \theta_2)$$
$$-m_2gr_2 \sin\theta_2 = m_2r_2^2\ddot{\theta}_2 + m_2r_1r_2\ddot{\theta}_1 \cos(\theta_1 - \theta_2)$$
$$- m_2r_1r_2\dot{\theta}_1^2 \sin(\theta_1 - \theta_2)$$

(a) Choose $x_1 = \theta_1, x_2 = \dot{\theta}_1, x_3 = \theta_2, x_4 = \dot{\theta}_2$, show that $[0; 0; 0; 0]$ is an equilibrium of the system.

(b) Show that the second-order Adam's-Bashforth method is given by:

$$x_{n+1} = x_n + h\left\{\frac{3}{2}f(x_n, t_n) - \frac{1}{2}f(x_{n-1}, t_{n-1})\right\}$$

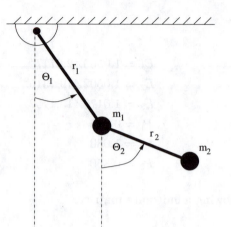

FIGURE 5.20
Double pendulum system

with

$$\epsilon_T = \frac{5}{12}\hat{x}^{(3)}(\tau)h^3$$

(c) Show that the third-order Adam's-Bashforth method is given by:

$$x_{n+1} = x_n + h\left\{\frac{23}{12}f(x_n, t_n) - \frac{16}{12}f(x_{n-1}, t_{n-1}) + \frac{5}{12}f(x_{n-2}, t_{n-2})\right\}$$

with

$$\epsilon_T = \frac{3}{8}\hat{x}^{(4)}(\tau)h^4$$

(d) Let $r_1 = 1, r_2 = 0.5$, and $g = 10$. Using the second-order Adam's-Bashforth method with $h=0.005$, plot the behavior of the system for an initial displacement of $\theta_1 = 25°$ and $\theta_2 = 10°$, for $T \in [0, 10]$ with

   i. $m_1 = 10$ and $m_2 = 10$.

   ii. $m_1 = 10$ and $m_2 = 5$.

   iii. $m_1 = 10$ and $m_2 = 1$.

(e) Repeat (d) using a variable step method with an upper LTE bound of $1e - 5$. Plot h versus t for each case. Discuss your solution.

(f) Repeat (d) and (e) using a third-order Adam's-Bashforth method.

Chapter 6

Optimization

The basic objective of any optimization method is to find the values of the
system state variables and/or parameters that minimize some cost function
of the system. The types of cost functions are system dependent and can
vary widely from application to application and are not necessarily strictly
measured in terms of dollars. Examples of engineering optimizations can
range from minimizing

- the error between a set of measured and calculated data,

- active power losses,

- the weight of a set of components that comprise the system,

- particulate output (emissions),

- system energy, or

- the distance between actual and desired operating points

to name a few possibilities. The basic formulation of any optimization can be
represented as minimizing a defined cost function subject to any physical or
operational constraints of the system:

$$\text{minimize } f(x, u) \quad x \in R^n \qquad (6.1)$$
$$u \in R^m$$

subject to

$$g(x, u) = 0 \text{ equality constraints} \qquad (6.2)$$
$$h(x, u) = 0 \text{ inequality constraints} \qquad (6.3)$$

where x is the vector of system states and u is the vector of system parameters. The basic approach is to find the vector of system parameters that
when substituted into the system model will result in the state vector x that
minimizes the cost function $f(x, u)$.

167

6.1   Least Squares State Estimation

In many physical systems, the system operating condition cannot be deter-mined directly by an analytical solution of known equations using a given set of known, dependable quantities. More frequently, the system operating con-dition is determined by the measurement of system states at different points throughout the system. In many systems, more measurements are made than are necessary to uniquely determine the operating point. This redundancy is often purposely designed into the system to counteract the effect of inac-curate or missing data due to instrument failure. Conversely, not all of the states may be available for measurement. High temperatures, moving parts, or inhospitable conditions may make it difficult, dangerous, or expensive to measure certain system states. In this case, the missing states must be esti-mated from the rest of the measured information of the system. This process is often known as *state estimation* and is the process of estimating unknown states from measured quantities. State estimation gives the "best estimate" of the state of the system in spite of uncertain, redundant, and/or conflicting measurements. A good state estimation will smooth out small random errors in measurements, detect and identify large measurement errors, and compen-sate for missing data. This process strives to minimize the error between the (unknown) true operating state of the system and the measured states.

The set of measured quantities can be denoted by the vector z which may include measurements of system states (such as voltage and current) or quan-tities that are functions of system states (such as power flows). Thus,

$$z^{true} = Ax \qquad (6.4)$$

where x is the set of system states and A is usually not square. The er-ror vector is the difference between the measured quantities z and the true quantities:

$$e = z - z^{true} = z - Ax \qquad (6.5)$$

Typically the minimum of the square of the error is desired to negate any effects of sign differences between the measured and true values. Thus, a state estimator endeavors to find the minimum of the squared error, or a *least squares minimization*:

$$\text{minimize } \|e\|^2 = e^T \cdot e = \sum_{i=1}^{m} \left[z_i - \sum_{j=1}^{m} a_{ij} x_j \right]^2 \qquad (6.6)$$

The squared error function can be denoted by $U(x)$ and is given by:

$$U(x) = e^T \cdot e = (z - Ax)^T (z - Ax) \qquad (6.7)$$

$$= \left(z^T - x^T A^T \right) (z - Ax) \qquad (6.8)$$

$$= z^T z - z^T Ax - x^T A^T z + x^T A^T Ax \qquad (6.9)$$

Note that the product $z^T A x$ is a scalar and so it can be equivalently written as

$$z^T A x = \left(z^T A x \right)^T = x^T A^T z$$

Therefore the squared error function is given by

$$U(x) = z^T z - 2x^T A^T z + x^T A^T A x \qquad (6.10)$$

The minimum of the squared error function can be found by an unconstrained optimization where the derivative of the function with respect to the states x is set to zero:

$$\frac{\partial U(x)}{\partial x} = 0 = -2A^T z + 2A^T A x \qquad (6.11)$$

Thus,

$$A^T A x = A^T z \qquad (6.12)$$

Thus, if $b = A^T z$ and $\hat{A} = A^T A$, then

$$\hat{A} x = b \qquad (6.13)$$

which can be solved by LU factorization. This state vector x is the best estimate (in the squared error) to the system operating condition from which the measurements z were taken. The measurement error is given by

$$e = z^{meas} - Ax \qquad (6.14)$$

Example 6.1
A set of measurements for the circuit shown in Figure 6.1 is given by:

Ammeter 1	z_1	4.27 A
Ammeter 2	z_2	-1.71 A
Voltmeter 1	z_3	3.47 V
Voltmeter 2	z_4	2.50 V

where $R_1 = R_3 = R_5 = 1.5\Omega$ and $R_2 = R_4 = 1.0\Omega$. Find the node voltages V_1 and V_2.

Solution 6.1 The Kirchoff voltage and current law equations for this system can be written as:

$$-V_1 + R_1 z_1 + z_3 = 0$$
$$-V_2 - R_5 z_2 + z_4 = 0$$
$$z_3/R_2 - z_1 + (z_3 - z_4)/R_3 = 0$$
$$z_4/R_4 + z_2 + (z_4 - z_3)/R_3 = 0$$

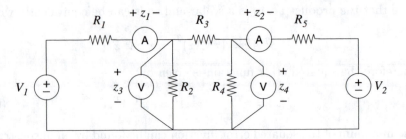

FIGURE 6.1
Circuit for Example 6.1

These equations can be rewritten in matrix form as:

$$\begin{bmatrix} R_1 & 0 & 1 & 0 \\ 0 & -R_5 & 0 & 1 \\ 1 & 0 & -\left(\frac{1}{R_2}+\frac{1}{R_3}\right) & \frac{1}{R_3} \\ 0 & 1 & -\frac{1}{R_3} & \frac{1}{R_3}+\frac{1}{R_4} \end{bmatrix} \begin{bmatrix} z_1 \\ z_2 \\ z_3 \\ z_4 \end{bmatrix} = \begin{bmatrix} 1 & 0 \\ 0 & 1 \\ 0 & 0 \\ 0 & 0 \end{bmatrix} \begin{bmatrix} V_1 \\ V_2 \end{bmatrix} \qquad (6.15)$$

To find the relationship between the measurements z and x, this equation must be reformulated as $z = Ax$. Note that this equation can be solved easily by LU factorization by considering each column of A individually. Thus,

$$\begin{bmatrix} R_1 & 0 & 1 & 0 \\ 0 & -R_5 & 0 & 1 \\ 1 & 0 & -\left(\frac{1}{R_2}+\frac{1}{R_3}\right) & \frac{1}{R_3} \\ 0 & 1 & -\frac{1}{R_3} & \frac{1}{R_3}+\frac{1}{R_4} \end{bmatrix} [A(:,1)] = \begin{bmatrix} 1 \\ 0 \\ 0 \\ 0 \end{bmatrix} \qquad (6.16)$$

Similarly,

$$\begin{bmatrix} R_1 & 0 & 1 & 0 \\ 0 & -R_5 & 0 & 1 \\ 1 & 0 & -\left(\frac{1}{R_2}+\frac{1}{R_3}\right) & \frac{1}{R_3} \\ 0 & 1 & -\frac{1}{R_3} & \frac{1}{R_3}+\frac{1}{R_4} \end{bmatrix} [A(:,2)] = \begin{bmatrix} 0 \\ 1 \\ 0 \\ 0 \end{bmatrix} \qquad (6.17)$$

yielding

$$\begin{bmatrix} z_1 \\ z_2 \\ z_3 \\ z_4 \end{bmatrix} = \begin{bmatrix} 0.4593 & -0.0593 \\ 0.0593 & -0.4593 \\ 0.3111 & 0.0889 \\ 0.0889 & 0.3111 \end{bmatrix} \begin{bmatrix} V_1 \\ V_2 \end{bmatrix} \qquad (6.18)$$

Thus,

$$b = A^T z = \begin{bmatrix} 0.4593 & 0.0593 & 0.3111 & 0.0889 \\ -0.0593 & -0.4593 & 0.0889 & 0.3111 \end{bmatrix} \begin{bmatrix} 4.27 \\ -1.71 \\ 3.47 \\ 2.50 \end{bmatrix}$$

$$= \begin{bmatrix} 3.1615 \\ 1.6185 \end{bmatrix}$$

and

$$\hat{A} = A^T A = \begin{bmatrix} 0.3191 & 0.0009 \\ 0.0009 & 0.3191 \end{bmatrix}$$

Leading to:

$$\begin{bmatrix} 0.3191 & 0.0009 \\ 0.0009 & 0.3191 \end{bmatrix} \begin{bmatrix} V_1 \\ V_2 \end{bmatrix} = \begin{bmatrix} 3.1615 \\ 1.6185 \end{bmatrix} \tag{6.19}$$

Solving this equation yields

$$V_1 = 9.8929$$
$$V_2 = 5.0446$$

The error between the measured values and the estimated values of this system is given by

$$e = z - Ax \tag{6.20}$$

$$= \begin{bmatrix} 4.27 \\ -1.71 \\ 3.47 \\ 2.50 \end{bmatrix} - \begin{bmatrix} 0.4593 & -0.0593 \\ 0.0593 & -0.4593 \\ 0.3111 & 0.0889 \\ 0.0889 & 0.3111 \end{bmatrix} \begin{bmatrix} 9.8929 \\ 5.0446 \end{bmatrix}$$

$$= \begin{bmatrix} 0.0255 \\ 0.0205 \\ -0.0562 \\ -0.0512 \end{bmatrix} \quad \blacksquare$$

6.1.1   Weighted Least Squares Estimation

If all measurements are treated equally in the least squares solution, then the less accurate measurements will affect the estimation as significantly as the more accurate measurements. As a result, the final estimation may contain large errors due to the influence of inaccurate measurements. By introducing a weighting matrix to emphasize the more accurate measurements more heavily than the less accurate measurements, the estimation procedure can then force the results to coincide more closely with the measurements of greater accuracy. This leads to the weighted least squares estimation:

$$\text{minimize } \|e\|^2 = e^T \cdot e = \sum_{i=1}^{m} w_i \left[z_i - \sum_{j=1}^{m} a_{ij} x_j \right]^2 \tag{6.21}$$

where w_i is a weighting factor reflecting the level of confidence in the measurement z_i.

Example 6.2

Suppose that the ammeters are known to have been more recently calibrated than the voltmeters; thus, the level of confidence in the current measurements is greater than the voltage measurements. Using the following weighting matrix, find the node voltages V_1 and V_2.

$$W = \begin{bmatrix} 100 & 0 & 0 & 0 \\ 0 & 100 & 0 & 0 \\ 0 & 0 & 50 & 0 \\ 0 & 0 & 0 & 50 \end{bmatrix}$$

Solution 6.2 By introducing the weighting matrix, the new minimum is given by

$$A^T W A x = A^T W z \tag{6.22}$$

The matrix $A^T W A$ is also known as the *gain* matrix. Using the same procedure as before, the weighted node voltage values are given by:

$$\begin{bmatrix} V_1 \\ V_2 \end{bmatrix} = \begin{bmatrix} 9.9153 \\ 5.0263 \end{bmatrix} \tag{6.23}$$

and the error vector is given by

$$e = \begin{bmatrix} 0.0141 \\ 0.0108 \\ -0.0616 \\ 0.0549 \end{bmatrix} \quad \blacksquare \tag{6.24}$$

Note that the added confidence in the current measurements has decreased the estimation error in the current, but the voltage measurement error is approximately the same.

Example 6.2 illustrates the impact of confidence weighting on the accuracy of the estimation. All instruments add some degree of error to the measured values, but the problem is how to quantify this error and account for it during the estimation process. In general, it can be assumed that the introduced errors have normal (Gaussian) distribution with zero mean and that each measurement is independent of all other measurements. This means that each measurement error is as likely to be greater than the true value as it is to be less than the true value. A zero mean Gaussian distribution has several attributes. The standard deviation of a zero mean Gaussian distribution is denoted by σ. This means that 68% of all measurements will fall within $\pm\sigma$ of the expected value, which is zero in a zero mean distribution. Further, 95% of all measurements will fall within $\pm2\sigma$ and 99% of all measurements will fall within $\pm3\sigma$. The variance of the measurement distribution is given by σ^2. This implies that if the variance of the measurements is relatively small, then the majority of measurements are close to the mean. One interpretation of this is that accurate measurements lead to small variance in the distribution.

This relationship between accuracy and variance leads to a straightforward approach from which to develop a weighting matrix for the estimation. Consider the squared error matrix given by

$$e \cdot e^T = \begin{bmatrix} e_1 \\ e_2 \\ e_3 \\ \vdots \\ e_m \end{bmatrix} \begin{bmatrix} e_1 & e_2 & e_3 & \cdots & e_m \end{bmatrix} \tag{6.25}$$

$$= \begin{bmatrix} e_1^2 & e_1 e_2 & e_1 e_3 & \cdots & e_1 e_m \\ e_2 e_1 & e_2^2 & e_2 e_3 & \cdots & e_2 e_m \\ \vdots & \vdots & \vdots & \vdots & \vdots \\ e_m e_1 & e_m e_2 & e_m e_3 & \cdots & e_m^2 \end{bmatrix} \tag{6.26}$$

where each e_i is the error in the ith measurement. The expected, or mean, value of each error product is given by $E[\cdot]$. The expected value of each of the diagonal terms is the variance of the ith error distribution σ_i^2. The expected value of each of the off-diagonal terms, or covariance, is zero because each measurement is assumed to be independent of every other measurement. Therefore, the expected value of the squared error matrix (also known as the covariance matrix) is

$$E\left[e \cdot e^T\right] = \begin{bmatrix} E\left[e_1^2\right] & E\left[e_1 e_2\right] & \cdots & E\left[e_1 e_m\right] \\ E\left[e_2 e_1\right] & E\left[e_2^2\right] & \cdots & E\left[e_2 e_m\right] \\ \vdots & \vdots & \vdots & \vdots \\ E\left[e_m e_1\right] & E\left[e_m e_2\right] & \cdots & E\left[e_m^2\right] \end{bmatrix} \tag{6.27}$$

$$= \begin{bmatrix} \sigma_1^2 & 0 & \cdots & 0 \\ 0 & \sigma_2^2 & \cdots & 0 \\ \vdots & \vdots & \vdots & \vdots \\ 0 & 0 & \cdots & \sigma_m^2 \end{bmatrix} \tag{6.28}$$

$$= R \tag{6.29}$$

With measurements taken from a particular meter, the smaller the variance of the measurements (i.e., the more consistent they are), the greater the level of confidence in that set of measurements. A set of measurements that have a high level of confidence should have a higher weighting than a set of measurements that have a larger variance (and therefore less confidence). Therefore, a plausible weighting matrix that reflects the level of confidence in each measurement set is the inverse of the covariance matrix $W = R^{-1}$. Thus, measurements that come from instruments with good consistency (small variance) will carry greater weight than measurements that come from less accurate instruments (high variance). Thus, one possible weighting matrix is given by

$$W = R^{-1} = \begin{bmatrix} \frac{1}{\sigma_1^2} & 0 & \cdots & 0 \\ 0 & \frac{1}{\sigma_2^2} & \cdots & 0 \\ \vdots & \vdots & \vdots & \vdots \\ 0 & 0 & \cdots & \frac{1}{\sigma_m^2} \end{bmatrix} \qquad (6.30)$$

6.1.2   Bad Data Detection

Frequently, a set of measurement will contain one or more data points from faulty or poorly calibrated instruments. Telemetered measurements are subject to noise or error in metering and communication. These "bad" data points typically fall outside of the standard deviation of the measurements and may affect the reliability of the state estimation process. In severe cases, the bad data may actually lead to grossly inaccurate results. Bad data may cause the accuracy of the estimate to deteriorate because of the "smearing" effect as the bad data will pull, or smear, the estimated values away from the true values. Therefore, it is desirable to develop a measure of the "goodness" of the data upon which the estimation is based. If the data lead to a good estimate of the states, then the error between the measured and calculated values will be small in some sense. If the error is large, then the data contain at least one bad data point. One error that is useful to consider is the estimated measurement error $\hat{e}$. This error is the difference between the actual measurements z and the estimated measurements $\hat{z}$. Recall the error vector from equation (6.14) where $e = z - Ax$; then the estimated measurement error becomes

$$\hat{e} = z - \hat{z} \qquad (6.31)$$
$$= z - A\hat{x} \qquad (6.32)$$
$$= z - A\left(A^T W A\right)^{-1} A^T W z \qquad (6.33)$$
$$= \left(I - A\left(A^T W A\right)^{-1} A^T W\right) z \qquad (6.34)$$
$$= \left(I - A\left(A^T W A\right)^{-1} A^T W\right)(e + Ax) \qquad (6.35)$$
$$= \left(I - A\left(A^T W A\right)^{-1} A^T W\right) e + A\left(I - \left(A^T W A\right)^{-1} A^T W A\right) x \qquad (6.36)$$
$$= \left(I - A\left(A^T W A\right)^{-1} A^T W\right) e \qquad (6.37)$$

Thus the variance of $\hat{e}$ can be calculated from

$$\hat{e}\hat{e}^T = (z - \hat{z})(z - \hat{z})^T \qquad (6.38)$$
$$= \left[I - A\left(A^T W A\right)^{-1} A^T W\right] ee^T \left[I - W A\left(A^T W A\right)^{-1} A^T\right] \qquad (6.39)$$

The expected, or mean, value of $\hat{e}\hat{e}^T$ is given by

$$E\left[\hat{e}\hat{e}^T\right] = \left[I - A\left(A^TWA\right)^{-1}A^TW\right]E\left[ee^T\right]\left[I - WA\left(A^TWA\right)^{-1}A^T\right]$$
(6.40)

Recall that $E\left[ee^T\right]$ is just the covariance matrix $R = W^{-1}$, which is a diagonal matrix. Thus

$$E\left[\hat{e}\hat{e}^T\right] = \left[I - A\left(A^TWA\right)^{-1}A^TW\right]\left[I - A\left(A^TWA\right)^{-1}A^TW\right]R \quad (6.41)$$

The matrix

$$\left[I - A\left(A^TWA\right)^{-1}A^TW\right]$$

has the unusual property that it is an *idempotent* matrix. An idempotent matrix M has the property that $M^2 = M$; thus no matter how many times M is multiplied by itself, it will still return the product M. Therefore,

$$E\left[\hat{e}\hat{e}^T\right] = \left[I - A\left(A^TWA\right)^{-1}A^TW\right]\left[I - A\left(A^TWA\right)^{-1}A^TW\right]R(6.42)$$

$$= \left[I - A\left(A^TWA\right)^{-1}A^TW\right]R \quad (6.43)$$

$$= R - A\left(A^TWA\right)^{-1}A^T \quad (6.44)$$

$$= R' \quad (6.45)$$

To determine whether the estimated values differ significantly from the measured values, a useful statistical measure is the χ^2 (chi-squared) test of inequality. This measure is based on the χ^2 probability distribution which differs in shape depending on its degrees of freedom k, which is the difference between the number of measurements and the number of states. By comparing the weighted sum of errors with the χ^2 value for a particular degree of freedom and significance level, it can be determined whether the errors exceed the bounds of what would be expected by chance alone. A significance level indicates the level of probability that the measurements are erroneous. A significance level of 0.05 indicates there is a 5% likelihood that bad data exist, or conversely, a 95% level of confidence in the goodness of the data. For example, for $k = 2$ and a significance level $\alpha = 0.05$, if the weighted sum of errors does not exceed a χ^2 of 5.99, then the set of measurements can be assured of being good with 95% confidence; otherwise the data must be rejected as containing at least one bad data point. Although the χ^2 test is effective in signifying the presence of bad data, it cannot identify locations. The identification of bad data locations continues to be an open research topic.

		χ^2 Values		
		α		
k	0.10	0.05	0.01	0.001
1	2.71	3.84	5.02	10.83
2	4.61	5.99	7.38	13.82
3	6.25	7.82	9.35	16.27
4	7.78	9.49	11.14	18.47
5	9.24	11.07	12.83	20.52
6	10.65	12.59	14.45	22.46
7	12.02	14.07	16.01	24.32
8	13.36	15.51	17.54	26.13
9	14.68	16.92	19.02	27.88
10	15.99	18.31	20.48	29.59
11	17.28	19.68	21.92	31.26
12	18.55	21.03	23.34	32.91
13	19.81	22.36	24.74	34.53
14	21.06	23.69	26.12	36.12
15	22.31	25.00	27.49	37.70
16	23.54	26.30	28.85	39.25
17	24.77	27.59	30.19	40.79
18	25.99	28.87	31.53	42.31
19	27.20	30.14	32.85	43.82
20	28.41	31.41	34.17	45.32
21	29.62	32.67	38.93	46.80
22	30.81	33.92	40.29	48.27
23	32.00	35.17	41.64	49.73
24	33.20	36.42	42.98	51.18
25	34.38	37.65	44.31	52.62
26	35.56	38.89	45.64	54.05
27	36.74	40.11	46.96	55.48
28	37.92	41.34	48.28	56.89
29	39.09	42.56	49.59	58.30
30	40.26	43.77	50.89	59.70

A test procedure to test for the existence of bad data is given by:

Test Procedure for Bad Data

1. Use z to estimate x

2. Calculate the error

$$e = z - Ax$$

3. Evaluate the weighted sum of squares

$$f = \sum_{i=1}^{m} \frac{1}{\sigma_i}^2 e_i^2$$

4. For $k = m - n$ and a specified probability α, if $f < \chi_{k,\alpha}^2$ then the data are good; otherwise at least one bad data point exists.

Example 6.3
Using the chi-square test of inequality with $\alpha = 0.01$, check for the presence of bad data in the measurements of Example 6.1.

Solution 6.3 The number of states in Example 6.1 is 2 and the number of measurements is 4; therefore $k = 4 - 2 = 2$. The weighted sum of squares is given by

$$f = \sum_{i=1}^{m=4} \frac{1}{\sigma_i} e_i^2$$
$$= 100(0.0141)^2 + 100(0.0108)^2 + 50(-0.0616)^2 + 50(0.0549)$$
$$= 0.3720$$

From the table of chi-squared values, the chi-square value for this example is 9.21. The weighted least squares error is less than the chi-square value; thus, this indicates that the estimated values are good to a confidence level of 99%. ∎

6.1.3 Nonlinear Least Squares State Estimation

As in the linear least squares estimation, the nonlinear least squares estimation attempts to minimize the square of the errors between a known set of measurements and a set of weighted nonlinear functions:

$$\text{minimize } f = \|e\|^2 = e^T \cdot e = \sum_{i=1}^{m} \frac{1}{\sigma^2} [z_i - h_i(x)]^2 \tag{6.46}$$

where $x \in R^n$ is the vector of unknowns to be estimated, $z \in R^m$ is the vector of measurements, σ_i^2 is the variance of the ith measurement, and $h(x)$ is the function vector relating x to z, where the measurement vector z can be a set of geographically distributed measurements, such as voltages and power flows.

In state estimation, the unknowns in the nonlinear equations are the state variables of the system. The state values that minimize the error are found by setting the derivatives of the error function to zero:

$$F(x) = H_x^T R^{-1} [z - h(x)] = 0 \qquad (6.47)$$

where

$$H_x = \begin{bmatrix} \frac{\partial h_1}{\partial x_1} & \frac{\partial h_1}{\partial x_2} & \cdots & \frac{\partial h_1}{\partial x_n} \\ \frac{\partial h_2}{\partial x_1} & \frac{\partial h_2}{\partial x_2} & \cdots & \frac{\partial h_2}{\partial x_n} \\ \vdots & \vdots & \vdots & \vdots \\ \frac{\partial h_m}{\partial x_1} & \frac{\partial h_m}{\partial x_2} & \cdots & \frac{\partial h_m}{\partial x_n} \end{bmatrix} \qquad (6.48)$$

and R is the matrix of measurement variances. Note that equation (6.47) is a set of nonlinear equations that must be solved using Newton-Raphson or another iterative numerical solver. In this case, the Jacobian of $F(x)$ is

$$J_F(x) = H_x^T(x) R^{-1} \frac{\partial}{\partial x} [z - h(x)] \qquad (6.49)$$

$$= -H_x^T(x) R^{-1} H_x(x) \qquad (6.50)$$

and the Newton-Raphson iteration becomes

$$\left[H_x^T \left(x^k \right) R^{-1} H_x \left(x^k \right) \right] \left[x^{k-1} - x^k \right] = H_x^T \left(x^k \right) R^{-1} \left[z - h(x^k) \right] \qquad (6.51)$$

which is solved repeatedly using LU factorization. At convergence, x^{k+1} is equal to the set of states that minimize the error function f of equation (6.46). The test procedure for bad data is the same as that for the linear state estimation.

6.2 Steepest Descent Algorithm

In an unconstrained system, the usual approach to minimizing the cost function is to set the function derivatives to zero and then solve for the system states from the set of resulting equations. In the majority of applications, however, the system states at the unconstrained minimum will not satisfy the constraint equations. Thus, an alternate approach is required to find the constrained minimum. One approach is to introduce an additional set of parameters λ, frequently known as *Lagrange multipliers*, to impose the constraints on the cost function. The augmented cost function then becomes

$$\text{minimize } f(x, u) - \lambda g(x, u) \qquad (6.52)$$

The augmented function in equation (6.52) can then be minimized by solving for the set of states that result from setting the derivatives of the augmented function to zero. Note that the derivative of equation (6.52) with respect to λ effectively enforces the equality constraint of equation (6.2).

Example 6.4
Minimize

$$C: \quad \frac{1}{2}\left(x^2 + y^2\right) \qquad (6.53)$$

subject to the following constraint:

$$2x - y = 5$$

Solution 6.4 Note that the function to be minimized is the equation for a circle. The unconstrained minimum of this function is the point at the origin with $x = 0$ and $y = 0$ which defines a circle with a radius of zero length. However, the circle must also intersect the line defined by the constraint equation; thus, the constrained circle must have a non-zero radius. The augmented cost function becomes

$$C^*: \quad \frac{1}{2}\left(x^2 + y^2\right) - \lambda\left(2x - y - 5\right) \qquad (6.54)$$

where λ represents the Lagrange multiplier. Setting the derivatives of the augmented cost function to zero yields the following set of equations:

$$0 = \frac{\partial C^*}{\partial x} = x - 2\lambda$$

$$0 = \frac{\partial C^*}{\partial y} = y + \lambda$$

$$0 = \frac{\partial C^*}{\partial \lambda} = 2x - y - 5$$

Solving this set of equations yields $[x \ y \ \lambda]^T = [2 \ -1 \ 1]^T$. The cost function of equation (6.53) evaluated at the minimum of the augmented cost function is:

$$C: \quad \frac{1}{2}\left((2)^2 + (-1)^2\right) = \frac{5}{2} \quad \blacksquare$$

If there is more than one equality constraint (i.e., if $g(x, u)$ of equation (6.2) is a vector of functions) then λ becomes a vector of multipliers and the augmented cost function becomes:

$$C^*: \quad f(x, u) - [\lambda]^T g(x, u) \qquad (6.55)$$

where the derivatives of C^* become:

$$\left[\frac{\partial C^*}{\partial \lambda}\right] = 0 = g(x, u) \tag{6.56}$$

$$\left[\frac{\partial C^*}{\partial x}\right] = 0 = \left[\frac{\partial f}{\partial x}\right] - \left[\frac{\partial g}{\partial x}\right]^T [\lambda] \tag{6.57}$$

$$\left[\frac{\partial C^*}{\partial u}\right] = 0 = \left[\frac{\partial f}{\partial u}\right] - \left[\frac{\partial g}{\partial u}\right]^T [\lambda] \tag{6.58}$$

Note that for any *feasible* solution to the equality constraint, equation (6.56) is satisfied, but the feasible solution may not be the optimal solution which minimizes the cost function. In this case, $[\lambda]$ can be obtained from equation (6.57) and then only

$$\left[\frac{\partial C^*}{\partial u}\right] \neq 0$$

This vector can be used as a gradient vector $[\nabla C]$ which is orthogonal to the contour of constant values of the cost function C. Thus,

$$[\lambda] = \left[\left[\frac{\partial g}{\partial x}\right]^T\right]^{-1} \left[\frac{\partial f}{\partial x}\right] \tag{6.59}$$

which gives

$$\nabla C = \left[\frac{\partial C^*}{\partial u}\right] = \left[\frac{\partial f}{\partial u}\right] - \left[\frac{\partial g}{\partial u}\right]^T [\lambda] \tag{6.60}$$

$$= \left[\frac{\partial f}{\partial u}\right] - \left[\frac{\partial g}{\partial u}\right]^T \left[\left[\frac{\partial g}{\partial x}\right]^T\right]^{-1} \left[\frac{\partial f}{\partial x}\right] \tag{6.61}$$

This relationship provides the foundation of the optimization method known as the *steepest descent* algorithm.

Steepest Descent Algorithm

1. Let k=0. Guess an initial vector $u^k = u^0$.

2. Solve the (possibly nonlinear) system of equations (6.56) for a feasible solution x.

3. Calculate C^{k+1} and ∇C^{k+1} from equation (6.61). If $\|C^{k+1} - C^k\|$ is less than some pre-defined tolerance, stop.

4. Calculate the new vector $u^{k+1} = u^k - \gamma \nabla C$, where γ is a positive number which is the user-defined "stepsize" of the algorithm.

5. k=k+1. Go to step 2.

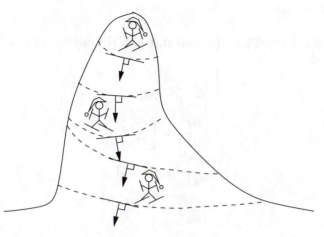

FIGURE 6.2
Example of steepest descent

In the steepest descent method, the u vector update direction is determined at each step of the algorithm by choosing the direction of the greatest change of the augmented cost function C^*. For example, consider a person skiing from the top of a mountain to the bottom as illustrated in Figure 6.2. The skier will travel in a straight path for a certain distance. At that point, he may no longer be pointed directly down the mountain. Therefore he will adjust his direction so that his skis point in the direction of steepest descent. The direction of steepest descent is perpendicular to the tangent of the curve of constant altitude (or cost). The distance the skier travels between adjustments is analogous to the stepsize γ of the algorithm. For small γ, the skier will be frequently altering direction; thus, his descent will be slow. For large γ, however, he may overshoot the foot of the mountain and will start ascending again. Thus the critical part of the steepest descent algorithm is the choice of γ. If γ is chosen small, then convergence to minimum value is more likely, but may require many iterations, whereas a large value of γ may result in oscillations about the minimum.

Example 6.5
Minimize

$$C: \quad x_1^2 + 2x_2^2 + u^2 = f(x_1, x_2, u) \tag{6.62}$$

subject to the following constraints:

$$0 = x_1^2 - 3x_2 + u - 3 \tag{6.63}$$
$$0 = x_1 + x_2 - 4u + 2 \tag{6.64}$$

Solution 6.5 To find ∇C of equation (6.61), the following partial derivatives are required:

$$\left[\frac{\partial f}{\partial u}\right] = 2u$$

$$\left[\frac{\partial f}{\partial x}\right] = \begin{bmatrix} 2x_1 \\ 4x_2 \end{bmatrix}$$

$$\left[\frac{\partial g}{\partial u}\right]^T = \begin{bmatrix} 1 & -4 \end{bmatrix}$$

$$\left[\frac{\partial g}{\partial x}\right] = \begin{bmatrix} 2x_1 & -3 \\ 1 & 1 \end{bmatrix}$$

yielding:

$$\nabla C = \left[\frac{\partial f}{\partial u}\right] - \left[\frac{\partial g}{\partial u}\right]^T \left[\left[\frac{\partial g}{\partial x}\right]^T\right]^{-1} \left[\frac{\partial f}{\partial x}\right]$$

$$= 2u - \begin{bmatrix} 1 & -4 \end{bmatrix} \left[\begin{bmatrix} 2x_1 & -3 \\ 1 & 1 \end{bmatrix}^T\right]^{-1} \begin{bmatrix} 2x_1 \\ 4x_2 \end{bmatrix}$$

Iteration 1

Let $u = 1, \gamma = 0.05$, and choose a stopping criterion of $\epsilon = 0.0001$. Solving for x_1 and x_2 yields two values for each with a corresponding cost function:

$$x_1 = 1.7016 \quad x_2 = 0.2984 \ f = 4.0734$$
$$x_1 = -4.7016 \ x_2 = 6.7016 \ f = 23.2828$$

The first set of values leads to the minimum cost function, so they are selected as the operating solution. Substituting $x_1 = 1.7016$ and $x_2 = 0.2984$ into the gradient function yields $\nabla C = 10.5705$ and the new value of u becomes:

$$u^{(2)} = u^{(1)} - \gamma \nabla C$$
$$= 1 - (0.05)(10.5705)$$
$$= 0.4715$$

Iteration 2

With $u = 0.4715$, solving for x_1 and x_2 again yields two values for each with a corresponding cost function:

$$x_1 = 0.6062 \quad x_2 = -0.7203 \ f = 1.6276$$
$$x_1 = -3.6062 \ x_2 = 3.4921 \quad f = 14.2650$$

The first set of values again leads to the minimum cost function, so they are selected as the operating solution. The difference in cost functions is

$$\left| C^{(1)} - C^{(2)} \right| = |4.0734 - 1.6276| = 2.4458$$

which is greater than the stopping criterion. Substituting these values into the gradient function yields $\nabla C = 0.1077$ and the new value of u becomes:

$$u^{(3)} = u^{(2)} - \gamma \nabla C$$
$$= 0.4715 - (0.05)(0.1077)$$
$$= 0.4661$$

Iteration 3

With $u = 0.4661$, solving for x_1 and x_2 again yields two values for each with a corresponding cost function:

$$x_1 = 0.5921 \quad x_2 = -0.7278 \ f = 1.6271$$
$$x_1 = -3.5921 \ x_2 = 3.4565 \quad f = 14.1799$$

The first set of values again leads to the minimum cost function, so they are selected as the operating solution. The difference in cost functions is

$$\left| C^{(2)} - C^{(3)} \right| = |1.6276 - 1.6271| = 0.0005$$

which is greater than the stopping criterion. Substituting these values into the gradient function yields $\nabla C = 0.0541$ and the new value of u becomes:

$$u^{(4)} = u^{(3)} - \gamma \nabla C$$
$$= 0.4661 - (0.05)(0.0541)$$
$$= 0.4634$$

Iteration 4

With $u = 0.4634$, solving for x_1 and x_2 again yields two values for each with a corresponding cost function:

$$x_1 = 0.5850 \quad x_2 = -0.7315 \ f = 1.6270$$
$$x_1 = -3.5850 \ x_2 = 3.4385 \quad f = 14.1370$$

The first set of values again leads to the minimum cost function, so they are selected as the operating solution. The difference in cost functions is

$$\left| C^{(3)} - C^{(4)} \right| = |1.6271 - 1.6270| = 0.0001$$

which satisfies the stopping criterion. Thus, the values $x_1 = 0.5850, x_2 = -0.7315$, and $u = 0.4634$ yield the minimum cost function $f = 1.6270$. ∎

6.3 Power System Applications

6.3.1 Optimal Power Flow

Many power system applications, such as the power flow, offer only a snap-shot of the system operation. Frequently, the system planner or operator is interested in the effect that making adjustments to the system parameters will have on the power flow through lines or system losses. Rather than making the adjustments in a random fashion, the system planner will attempt to optimize the adjustments according to some objective function. This objective function can be chosen to minimize generating costs, reservoir water levels, or system losses, among others. The optimal power flow problem is to formulate the power flow problem to find system voltages and generated powers within the framework of the objective function. In this application, the inputs to the power flow are systematically adjusted to maximize (or minimize) a scalar function of the power flow state variables. The two most common objective functions are minimization of generating costs and minimization of active power losses.

The time frame of optimal power flow is on the order of minutes to one hour; therefore it is assumed that the optimization occurs using only those units that are currently on-line. The problem of determining whether or not to engage a unit, at what time, and for how long is part of the *unit commitment* problem and is not covered here. The minimization of active transmission losses saves both generating costs and creates a higher generating reserve margin.

Usually generator cost curves (the curves that relate generated power to the cost of such generation) are given as piecewise linear incremental costs curves. This has its origin in the simplification of concave cost functions with the valve points as cost curve breakpoints [13]. Piecewise linear incremental cost curves correspond to piecewise quadratic cost curves by integrating the incremental cost curves. This type of objective function lends itself easily to the *economic dispatch*, or λ-dispatch problem where only generating units are considered in the optimization. In this process, system losses and constraints on voltages and line powers are neglected. This economic dispatch method is illustrated in the following example.

Example 6.6

Three generators with the following cost functions serve a load of 952 MW. Assuming a lossless system, calculate the optimal generation scheduling.

$$C_1 : \ P_1 + 0.0625P_1^2 \quad \$/hr$$
$$C_2 : \ P_2 + 0.0125P_2^2 \quad \$/hr$$
$$C_3 : \ P_3 + 0.0250P_3^2 \quad \$/hr$$

Solution 6.6 The first step in determining the optimal scheduling of the generators is to construct the problem in the general form. Thus the optimization statement is:

$$\text{Minimize C: } P_1 + 0.0625P_1^2 + P_2 + 0.0125P_2^2 + P_3 + 0.0250P_3^2$$
$$\text{Subject to: } P_1 + P_2 + P_3 - 952 = 0$$

From this statement, the constrained cost function becomes

$$C^* : \quad P_1 + 0.0625P_1^2 + P_2 + 0.0125P_2^2 + P_3 + 0.0250P_3^2 - \lambda\,(P_1 + P_2 + P_3 - 952) \tag{6.65}$$

Setting the derivatives of C^* to zero yields the following set of linear equations:

$$\begin{bmatrix} 0.125 & 0 & 0 & -1 \\ 0 & 0.025 & 0 & -1 \\ 0 & 0 & 0.050 & -1 \\ 1 & 1 & 1 & 0 \end{bmatrix} \begin{bmatrix} P_1 \\ P_2 \\ P_3 \\ \lambda \end{bmatrix} = \begin{bmatrix} -1 \\ -1 \\ -1 \\ 952 \end{bmatrix} \tag{6.66}$$

Solving equation (6.66) yields

$$P_1 = 112\ MW$$
$$P_2 = 560\ MW$$
$$P_3 = 280\ MW$$
$$\lambda = \$15/MW hr$$

for a constrained cost of \$7,616/hr. ∎

This is the generation scheduling that minimizes the hourly cost of production. The value of λ is the *incremental* or *break-even* cost of production. This gives a company a price cut-off for buying or selling generation: if they can purchase generation for less than λ, then their overall costs will decrease. Likewise, if generation can be sold for greater than λ, their overall costs will decrease. Also note that at the optimal scheduling:

$$\lambda = 1 + 0.125P_1 = 1 + 0.025P_2 = 1 + 0.050P_3 \tag{6.67}$$

Since λ is the incremental cost for the system, this point is also called the point of "equal incremental cost," and the generation schedule is said to satisfy the "equal incremental cost criterion." Any deviation in generation from the equal increment cost scheduling will result in an increase in the production cost C.

Example 6.7
If a buyer is willing to pay \$16/MW hr for generation, how much excess generation should be produced and sold, and what is the profit for this transaction?

Solution 6.7 From Example 6.6, the derivatives of the augmented cost function yield the following relationships between generation and λ:

$$P_1 = 8\,(\lambda - 1)$$
$$P_2 = 40\,(\lambda - 1)$$
$$P_3 = 20\,(\lambda - 1)$$

from which the equality constraint yields:

$$8\,(\lambda - 1) + 40\,(\lambda - 1) + 20\,(\lambda - 1) - 952 = 0 \qquad (6.68)$$

To determine the excess amount, the equality equation (6.68) will be augmented and then evaluated at $\lambda = \$16/\mathrm{MW}$ hr:

$$8\,(16 - 1) + 40\,(16 - 1) + 20\,(16 - 1) - (952 + \mathrm{excess}) = 0 \qquad (6.69)$$

Solving equation (6.69) yields a required excess of 68 MW, and $P_1 = 120$ MW, $P_2 = 600$ MW, and $P_3 = 300$ MW. The total cost of generation becomes

$$C: \quad P_1 + 0.0625P_1^2 + P_2 + 0.0125P_2^2 + P_3 + 0.0250P_3^2 = \$8,670/hr \quad (6.70)$$

but the amount recovered by the sale of generation is the amount of excess times the incremental cost λ,

$$68MW \times \$16/MWhr = \$1,088/hr$$

Therefore, the total cost is $\$8,670 - 1,088 = \$7,580/\mathrm{hr}$. This amount is $\$34/\mathrm{hr}$ less than the original cost of $\$7,616/\mathrm{hr}$; thus, $\$34/\mathrm{hr}$ is the profit achieved from the sale of the excess generation at $\$16/\mathrm{MW}$ hr. ∎

Figure 6.3 shows an incremental cost table for a medium size utility. The incremental cost of generation is listed vertically along the left hand side of the table. The various generating units are listed across the top from least expensive to most expensive (left to right). Nuclear units are among the least expensive units to operate and the nuclear unit *Washington* at the far left can produce up to 1222 MW at an incremental cost of 7.00 \$/MW hr. This incremental cost is half of the next least expensive unit *Adams* at 14 \$/MW hr which is a coal unit. As the available units become increasingly more expensive to operate, the incremental cost also increases.

Example 6.8
What is the incremental cost for the utility to produce 4500 MW?

Solution 6.8 To find the incremental cost that corresponds to 4500 MW from the Incremental Cost Table in Figure 6.3, the maximum generation available from each unit is summed until it equals 4500 MW. This amount is reached

Dollar per MWhr	WASH NUC 0.41	ADAMS 1-2 COAL 0.95	ADAMS 3-4 COAL 0.95	JEFF 1-2 COAL 0.93	MADI 1-2 COAL 0.96	MONR 4 COAL 1.41	MONROE 3 COAL 1.41	MONROE 3 GAS 2.25	MONROE 1-2 COAL 1.41	MONROE 1-2 GAS 2.25	Q ADAMS 1-2 OIL 3.65	Q ADAMS 1-2 GAS 2.25	Q ADAMS 3-4 OIL 3.65	Q ADAMS 3-4 GAS 2.25	QADA 5-6 OIL 3.65	JACK GAS 2.10	VBUR OIL 3.50	HARR OIL 3.50	TYLE OIL 3.55	POLK OIL 3.65
7.00	1222																			
14.00		160	240																	
14.50		240	310																	
15.00		320	380	220																
15.50		390	450	410																
16.00		470	520	590																
16.50		540	587	608	502		100		20											
17.00		587				110	130		30											
17.50						170	160		50											
18.00						230	200		60											
18.50						290	230		80											
19.00						360	260		90											
19.50							290		110											
20.00							298		130											
20.50									142											
22.00								90		20		20		20						
25.00										80		45		60						
28.00										142				95						
31.00																108				
34.00											30		20		30					
37.00											40		40		50					
40.00											45									
43.00																				
46.00													60		80					
49.00													80		95					
52.00													95					63		
55.00																	48		189	
64.00																				30

FIGURE 6.3
Incremental cost table

by the gas unit of *Monroe 1-2*. This corresponds to an incremental cost of 28.00 \$/MW hr. This is the breakeven point for 4500 MW.

If power can be purchased for less than 28.00 \$/MW hr, the utility should purchase generation. ∎

The primary drawback with the equal incremental cost scheduling is that it neglects all losses in the system. The only enforced equality constraint is that the sum of the generation must equal the total load demand. In reality, however, the sum of the generation must equal the load demand plus any system losses. In the consideration of system losses, the equality constraints must include the set of power flow equations, and the optimization process must to be extended to the steepest descent, or similar, approach [5].

Example 6.9
Consider the three machine system shown in Figure 6.4. This system has the same parameters as the three bus system of Example 3.6 except that bus 3 has been converted to a generator bus with a voltage magnitude of 1.0 pu. The total load and cost functions of the generators are the same as in Example 6.6. Using the equal cost criterion scheduling as a starting point, find the optimal scheduling of this system considering losses.

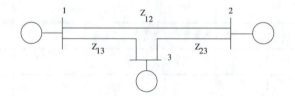

FIGURE 6.4
Figure for Example 6.9

Solution 6.9 Following the steepest descent procedure detailed in Section 6.2, the first step is to develop an expression for the gradient ∇C, where

$$\nabla C = \left[\frac{\partial f}{\partial u}\right] - \left[\frac{\partial g}{\partial u}\right]^T \left[\left[\frac{\partial g}{\partial x}\right]^T\right]^{-1} \left[\frac{\partial f}{\partial x}\right] \tag{6.71}$$

where f is the sum of the generator costs:

$$f: \quad C_1 + C_2 + C_3 = P_1 + 0.0625P_1^2 + P_2 + 0.0125P_2^2 + P_3 + 0.0250P_3^2$$

g is the set of load flow equations:

$$g_1: \quad 0 = P_2 - P_{L2} - V_2 \sum_{i=1}^{3} V_i Y_{2i} \cos(\delta_2 - \delta_i - \phi_{2i})$$

$$g_2: \quad 0 = P_3 - P_{L3} - V_3 \sum_{i=1}^{3} V_i Y_{3i} \cos(\delta_3 - \delta_i - \phi_{3i})$$

where P_{Li} denotes the active power load at bus i, the set of inputs u is the set of independent generation settings:

$$u = \begin{bmatrix} P_2 \\ P_3 \end{bmatrix}$$

and x is the set of unknown states

$$x = \begin{bmatrix} \delta_2 \\ \delta_3 \end{bmatrix}$$

The generator setting P_1 is not an input because it is the slack bus generation and cannot be independently set. From these designations, the various partial derivatives required for ∇C can be derived:

$$\left[\frac{\partial g}{\partial u}\right] = \begin{bmatrix} 1 \\ 1 \end{bmatrix} \tag{6.72}$$

$$\left[\frac{\partial g}{\partial x}\right] = \begin{bmatrix} \frac{\partial g_1}{\partial \delta_2} & \frac{\partial g_1}{\partial \delta_3} \\ \frac{\partial g_2}{\partial \delta_2} & \frac{\partial g_2}{\partial \delta_3} \end{bmatrix} \tag{6.73}$$

where

$$\frac{\partial g_1}{\partial \delta_2} = V_2 \left(V_1 Y_{12} \sin(\delta_2 - \delta_1 - \phi_{21}) + V_3 Y_{13} \sin(\delta_2 - \delta_3 - \phi_{23}) \right) \tag{6.74}$$

$$\frac{\partial g_1}{\partial \delta_3} = -V_2 V_3 Y_{32} \sin(\delta_2 - \delta_3 - \phi_{23}) \tag{6.75}$$

$$\frac{\partial g_2}{\partial \delta_2} = -V_3 V_2 Y_{23} \sin(\delta_3 - \delta_2 - \phi_{32}) \tag{6.76}$$

$$\frac{\partial g_2}{\partial \delta_3} = V_3 \left(V_1 Y_{13} \sin(\delta_3 - \delta_1 - \phi_{31}) + V_2 Y_{23} \sin(\delta_3 - \delta_2 - \phi_{32}) \right) \tag{6.77}$$

and

$$\left[\frac{\partial f}{\partial u}\right] = \begin{bmatrix} 1 + 0.025 P_2 \\ 1 + 0.050 P_3 \end{bmatrix} \tag{6.78}$$

Finding the partial derivative $\left[\frac{\partial f}{\partial x}\right]$ is slightly more difficult since the cost function is not written as a direct function of x. Recall, however, that P_1 is not an input, but is actually a quantity that depends on x, i.e.,

$$P_1 = V_1 \left(V_1 Y_{11} \cos(\delta_1 - \delta_1 - \phi_{11}) \right.$$
$$\left. + V_2 Y_{12} \cos(\delta_1 - \delta_2 - \phi_{12}) + V_3 Y_{13} \cos(\delta_1 - \delta_3 - \phi_{13}) \right) \tag{6.79}$$

Thus, using the chain rule,

$$\left[\frac{\partial f}{\partial x}\right] = \left[\frac{\partial f}{\partial P_1}\right]\left[\frac{\partial P_1}{\partial x}\right] \tag{6.80}$$

$$= (1 + 0.125P_1)\left[\begin{array}{c} V_1 V_2 Y_{12} \sin(\delta_1 - \delta_2 - \phi_{12}) \\ V_1 V_3 Y_{13} \sin(\delta_1 - \delta_3 - \phi_{13}) \end{array}\right] \tag{6.81}$$

From the previous example, the initial values of $P_2 = 0.56$ pu and $P_3 = 0.28$ pu are obtained from the equal incremental cost rule. Using $P_2 = 0.56$ pu and $P_3 = 0.28$ pu as inputs into the power flow yields the following states: $[\delta_2 \quad \delta_3] = [0.0286 \quad -0.0185]$ in radians and $P_1 = 0.1152$. Converting the generated powers to MW and substituting these values into the partial derivatives yields:

$$\left[\frac{\partial g}{\partial u}\right] = \left[\begin{array}{cc} 1 & 0 \\ 0 & 1 \end{array}\right] \tag{6.82}$$

$$\left[\frac{\partial g}{\partial x}\right] = \left[\begin{array}{cc} -13.3267 & 9.9366 \\ 9.8434 & -19.9219 \end{array}\right] \tag{6.83}$$

$$\left[\frac{\partial f}{\partial u}\right] = \left[\begin{array}{c} 15.0000 \\ 15.0000 \end{array}\right] \tag{6.84}$$

$$\left[\frac{\partial f}{\partial x}\right] = 15.4018\left[\begin{array}{c} -52.0136 \\ -155.8040 \end{array}\right] \tag{6.85}$$

which yields

$$\nabla C = \left[\begin{array}{c} -0.3256 \\ -0.4648 \end{array}\right] \tag{6.86}$$

Thus, the new values for the input generation are:

$$\left[\begin{array}{c} P_2 \\ P_3 \end{array}\right] = \left[\begin{array}{c} 560 \\ 280 \end{array}\right] - \gamma\left[\begin{array}{c} -0.3256 \\ -0.4648 \end{array}\right] \tag{6.87}$$

With $\gamma = 1$, the updated generation is $P_2 = 560.3$ and $P_3 = 280.5$ MW.

Already the gradient ∇C is very small, indicating that the generation values from the equal incremental cost process were relatively close to the optimal values, even considering losses. Proceeding with one more iteration yields the final generation values for all of the generators:

$$\left[\begin{array}{c} P_1 \\ P_2 \\ P_3 \end{array}\right] = \left[\begin{array}{c} 112.6 \\ 560.0 \\ 282.7 \end{array}\right] \quad \text{MW}$$

which yields a cost of \$7,664/MW hr. Note that this amount is greater than the calculated cost for the equal incremental cost function. This increase is due to the extra generation required to satisfy the losses in the system. ∎

Often the steepest descent method may indicate either states or inputs lie outside of their physical constraints. For example, the algorithm may result in a power generation value that exceeds the physical maximum output of the generating unit. Similarly, the resulting bus voltages may lie outside of the desired range (usually ± 10% of unity). These are violations of the *inequality constraints* of the problem. In these cases, the steepest descent algorithm must be modified to reflect these physical limitations. There are several approaches to account for limitations and these approaches depend on whether or not the limitation is on the input (independent) or on the state (dependent).

6.3.1.1   Limitations on Independent Variables

If the application of the steepest descent algorithm results in an updated value of input that exceeds the specified limit, then the most straightforward method of handling this violation is simply to set the input state equal to its limit and continue with the algorithm except with one less degree of freedom.

Example 6.10

Repeat Example 6.9 except that the generators must satisfy the following limitations:

$$80 \leq P_1 \leq 1200 \text{ MW}$$
$$450 \leq P_2 \leq 750 \text{ MW}$$
$$150 \leq P_3 \leq 250 \text{ MW}$$

Solution 6.10 From the solution of Example 6.9, the output of generator 3 exceeds the maximum limit of 0.25 pu. Therefore after the first iteration in the previous example, P_3 is set to 0.25 pu. The new partial derivatives become:

$$\left[\frac{\partial g}{\partial u}\right] = \begin{bmatrix} 0 \\ 1 \end{bmatrix} \tag{6.88}$$

$$\left[\frac{\partial g}{\partial x}\right] = \text{same} \tag{6.89}$$

$$\left[\frac{\partial f}{\partial u}\right] = [1 + 0.025P_2] \tag{6.90}$$

$$\left[\frac{\partial f}{\partial x}\right] = \text{same} \tag{6.91}$$

From the constrained steepest descent, the new values of generation become:

$$\begin{bmatrix} P_1 \\ P_2 \\ P_3 \end{bmatrix} = \begin{bmatrix} 117.1 \\ 588.3 \\ 250.0 \end{bmatrix} \text{ MW}$$

with a cost of \$7,703/MW hr which is higher than the unconstrained cost of generation of \$7,664/MW hr. As more constraints are added to the system, the system is moved away from the optimal operating point, increasing the cost of generation. ■

6.3.1.2   Limitations on Dependent Variables

In many cases, the physical limitations of the system are imposed upon states that are dependent variables in the system description. In this case, the inequality equations are functions of x and must be added to the cost function. Examples of limitations on dependent variables include maximum line flows or bus voltage levels. In these cases, the value of the states cannot be independently set, but must be enforced indirectly. One method of enforcing an inequality constraint is to introduce a *penalty function* into the cost function. A penalty function is a function that is small when the state is far away from its limit, but becomes increasingly larger the closer the state is to its limit. Typical penalty functions include:

$$p(h) = e^{kh} \quad k > 0 \tag{6.92}$$

$$p(h) = x^{2n}e^{kh} \quad n, k > 0 \tag{6.93}$$

$$p(h) = ax^{2n}e^{kh} + be^{kh} \quad n, k, a, b > 0 \tag{6.94}$$

and the cost function becomes

$$C^* : \quad C(u, x) + \lambda^T g(u, x) + p\left(h(u, x) - h^{max}\right) \tag{6.95}$$

This cost equation is then minimized in the usual fashion by setting the appropriate derivatives to zero. This method has the advantage of simplicity of implementation, but also has several disadvantages. The first disadvantage is that the choice of penalty function is often a heuristic choice and can vary by application. A second disadvantage is that this method cannot enforce *hard* limitations on states, i.e., the cost function becomes large if the maximum is exceeded, but the state is allowed to exceed its maximum. In many applications this is not a serious disadvantage. If the power flow on a transmission line slightly exceeds its maximum, it is reasonable to assume that the power system will continue to operate, at least for a finite length of time. If, however, the physical limit is the height above ground for an airplane, then even a slightly negative altitude will have dire consequences. Thus the use of penalty functions to enforce limits must be used with caution and is not applicable for all systems.

Example 6.11
Repeat Example 6.9, except use penalty functions to limit the power flow across line 2-3 to 0.4 per unit.

Solution 6.11 The power flow across line 2-3 in Example 6.9 is given by

$$P_{23} = V_2 V_3 Y_{23} \cos(\delta_2 - \delta_3 - \phi_{23}) - V_2^2 Y_{23} \cos \phi_{23} \qquad (6.96)$$
$$= 0.467 \text{ per unit}$$

If P_{23} exceeds 0.4 per unit, then the penalty function

$$p(h) = \left(1000 V_2 V_3 Y_{23} \cos(\delta_2 - \delta_3 - \phi_{23}) - 1000 V_2^2 Y_{23} \cos \phi_{23} - 400\right)^2 \quad (6.97)$$

will be appended to the cost function. The partial derivatives remain the same with the exception of $\left[\frac{\partial f}{\partial x}\right]$ which becomes:

$$\left[\frac{\partial f}{\partial x}\right] = \left[\frac{\partial f}{\partial P_1}\right]\left[\frac{\partial P_1}{\partial x}\right] + \left[\frac{\partial f}{\partial P_{23}}\right]\left[\frac{\partial P_{23}}{\partial x}\right] \qquad (6.98)$$

$$= (1 + 0.125 P_1) \begin{bmatrix} V_1 V_2 Y_{12} \sin(\delta_1 - \delta_2 - \phi_{1,2}) \\ V_1 V_3 Y_{13} \sin(\delta_1 - \delta_3 - \phi_{1,3}) \end{bmatrix}$$

$$+ 2(P_{23} - 400) \begin{bmatrix} -V_2 V_3 Y_{23} \sin(\delta_2 - \delta_3 - \phi_{23}) \\ V_2 V_3 Y_{23} \sin(\delta_2 - \delta_3 - \phi_{23}) \end{bmatrix} \qquad (6.99)$$

Proceeding with the steepest gradient algorithm iterations yields the final constrained optimal generation scheduling:

$$\begin{bmatrix} P_1 \\ P_2 \\ P_3 \end{bmatrix} = \begin{bmatrix} 128.5 \\ 476.2 \\ 349.9 \end{bmatrix} \text{ MW}$$

and $P_{23} = 400$ MW. The cost for this constrained scheduling is \$7,882/MW hr which is slightly greater than the non-constrained cost. ∎

In the case where hard limits must be imposed, an alternate approach to enforcing the inequality constraints must be employed. In this approach, the inequality constraints are added as additional equality constraints with the inequality set equal to the limit (upper or lower) that is violated. This in essence introduces an additional set of Lagrangian multipliers. This is often referred to as the dual variable approach, because each inequality has the potential of resulting in two equalities: one for the upper limit and one for the lower limit. However, the upper and lower limit cannot be simultaneously violated; thus, out of the possible set of additional Lagrangian multipliers only one of the two will be included at any given operating point and thus the dual limits are mutually exclusive.

Example 6.12
Repeat Example 6.11 using the dual variable approach.

Solution 6.12 By introducing the additional equation

$$P_{23} = V_2 V_3 Y_{23} \cos(\delta_2 - \delta_3 - \phi_{23}) - V_2^2 Y_{23} \cos \phi_{23} = 0.400 \text{ per unit} \quad (6.100)$$

to the equality constraints adds an additional equation to the set of $g(x)$. Therefore an additional unknown must be added to the state vector x to yield a solvable set of equations (three equations in three unknowns). Either P_{G2} or P_{G3} can be chosen as the additional unknown. In this example, P_{G3} will be chosen. The new system Jacobian becomes:

$$\left[\frac{\partial g}{\partial x}\right] = \begin{bmatrix} \frac{\partial g_1}{\partial x_1} & \frac{\partial g_1}{\partial x_2} & \frac{\partial g_1}{\partial x_3} \\ \frac{\partial g_2}{\partial x_1} & \frac{\partial g_2}{\partial x_2} & \frac{\partial g_2}{\partial x_3} \\ \frac{\partial g_3}{\partial x_1} & \frac{\partial g_3}{\partial x_2} & \frac{\partial g_3}{\partial x_3} \end{bmatrix} \qquad (6.101)$$

where

$$\frac{\partial g_1}{\partial x_1} = V_2 \left(V_1 Y_{12} \sin\left(\delta_2 - \delta_1 - \phi_{21}\right) + V_3 Y_{13} \sin\left(\delta_2 - \delta_3 - \phi_{23}\right)\right)$$

$$\frac{\partial g_1}{\partial x_2} = -V_2 V_3 Y_{32} \sin\left(\delta_2 - \delta_3 - \phi_{23}\right)$$

$$\frac{\partial g_1}{\partial x_3} = 0$$

$$\frac{\partial g_2}{\partial x_1} = -V_3 V_2 Y_{23} \sin\left(\delta_3 - \delta_2 - \phi_{32}\right)$$

$$\frac{\partial g_2}{\partial x_2} = V_3 V_1 Y_{13} \sin\left(\delta_3 - \delta_1 - \phi_{31}\right) + V_2 Y_{23} \sin\left(\delta_3 - \delta_2 - \phi_{32}\right)$$

$$\frac{\partial g_2}{\partial x_3} = 1$$

$$\frac{\partial g_3}{\partial x_1} = -V_2 V_3 Y_{23} \sin\left(\delta_2 - \delta_3 - \phi_{23}\right)$$

$$\frac{\partial g_3}{\partial x_2} = V_2 V_3 Y_{23} \sin\left(\delta_2 - \delta_3 - \phi_{23}\right)$$

$$\frac{\partial g_3}{\partial x_3} = 0$$

and

$$\left[\frac{\partial g}{\partial u}\right] = \begin{bmatrix} 1 \\ 0 \\ 0 \end{bmatrix}; \qquad \left[\frac{\partial f}{\partial u}\right] = [1 + 0.025 P_{G2}]$$

Similar to Example 6.9, the chain rule is used to obtain $\left[\frac{\partial f}{\partial x}\right]$:

$$\left[\frac{\partial f}{\partial x}\right] = \left[\frac{\partial C}{\partial P_{G1}}\right]\left[\frac{\partial P_{G1}}{\partial x}\right] + \left[\frac{\partial C}{\partial P_{G3}}\right]\left[\frac{\partial P_{G3}}{\partial x}\right] \qquad (6.102)$$

$$= (1 + 0.125 P_{G1}) \begin{bmatrix} V_1 V_2 Y_{12} \sin\left(\delta_1 - \delta_2 - \phi_{12}\right) \\ V_1 V_3 Y_{13} \sin\left(\delta_1 - \delta_3 - \phi_{13}\right) \\ 0 \end{bmatrix} + (1 + 0.050 P_{G3}) \times$$

$$\begin{bmatrix} V_3 V_2 Y_{32} \sin\left(\delta_3 - \delta_2 - \phi_{32}\right) \\ -V_3 \left(V_1 Y_{13} \sin\left(\delta_3 - \delta_1 - \phi_{31}\right) + V_2 Y_{23} \sin\left(\delta_3 - \delta_2 - \phi_{32}\right)\right) \\ 0 \end{bmatrix} \qquad (6.103)$$

Substituting these partial derivatives into the expression for ∇C of equation (6.71) yields the same generation scheduling as Example 6.11. ∎

6.3.2 State Estimation

In power system state estimation, the estimated variables are the voltage magnitudes and the voltage phase angles at the system buses. The input to the state estimator is the active and reactive powers of the system, measured either at the injection sites or on the transmission lines. The state estimator is designed to give the best estimates of the voltages and phase angles minimizing the effects of the measurement errors. Another consideration for the state estimator is to determine if a sufficient number of measurements are available to fully estimate the power system. This is the notion of system observability.

A set of specified measurements of a power system is said to be *observable* if the entire state vector of bus voltage magnitude and phase angles can be estimated from the set of available measurements. An unobservable system is one in which the set of measurements do not span the entire state space. The power system is observable if the matrix H_x in equation (6.47) has rank n (full rank), where the number of measurements m is greater than or equal to the number of system states n. A *redundant* measurement is one whose addition to the measurement does not increase the rank of the matrix H_x.

The observability of a power system can be determined by examining the measurement set and the structure of the power system. A *tree* is a set of measurements (either bus or line) that spans the entire set of power system buses. In other words, by graphically connecting the buses and lines that contribute to the set of measurements, the entire set of system buses can be connected by a single connected graph. A power system can be made observable by adding measurements at those lines that will connect disjoint trees.

Example 6.13

The SCADA system for the power network shown in Figure 6.5 reports the following measurements and variances:

z_i	state	measurement	variance (σ^2)
1	V_3	0.975	0.010
2	P_{13}	0.668	0.050
3	Q_{21}	-0.082	0.075
4	P_3	-1.181	0.050
5	Q_2	-0.086	0.075

Estimate the power system states and use the chi-square test of inequality with $\alpha = 0.01$ to check for the presence of bad data in the measurements.

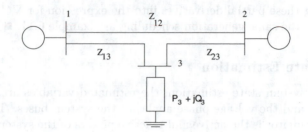

FIGURE 6.5
Example power system

Solution 6.13 The first step in the estimation process is to identify and enumerate the unknown states. In this example, the unknowns are $[x_1 \; x_2 \; x_3]^T = [\delta_2 \; \delta_3 \; V_3]^T$. After the states are identified, the next step in the estimation process is to identify the appropriate functions $h(x)$ that correspond to each of the measurements. The nonlinear function that is being driven to zero to minimize the weighted error is

$$F(x) = H_x^T R^{-1} [z - h(x)] = 0 \tag{6.104}$$

where the set of $z - h(x)$ is given by

$$z_1 - h_1(x) = V_3 - x_3$$
$$z_2 - h_2(x) = P_{13} - \left(V_1 x_3 Y_{13} \cos(-x_2 - \phi_{13}) - V_1^2 Y_{13} \cos \phi_{13}\right)$$
$$z_3 - h_3(x) = Q_{21} - \left(V_2 V_1 Y_{21} \sin(x_1 - \phi_{21}) + V_2^2 Y_{21} \sin \phi_{21}\right)$$
$$z_4 - h_4(x) = P_3 - \left(x_3 V_1 Y_{31} \cos(x_2 - \phi_{31}) + x_3 V_2 Y_{32} \cos(x_2 - x_1 - \phi_{32})\right)$$
$$\qquad\qquad + x_3^2 Y_{33} \cos \phi_{33})$$
$$z_5 - h_5(x) = Q_2 - \left(V_2 V_1 Y_{21} \sin(x_1 - \phi_{21}) - V_2^2 Y_{22} \sin \phi_{22}\right)$$
$$\qquad\qquad + V_2 x_3 Y_{23} \sin(x_1 - x_2 - \phi_{23}))$$

and the matrix of partial derivatives for the set of functions (6.104) is $H_x =:$

$$\begin{bmatrix} 0 & 0 & 1 \\ 0 & V_1 x_3 Y_{13} \sin(-x_2 - \phi_{13}) & V_1 Y_{13} \cos(-x_2 - \phi_{13}) \\ V_1 V_2 Y_{21} \cos(x_1 - \phi_{21}) & 0 & 0 \\ x_3 V_2 Y_{32} \sin(x_2 - x_1 - \phi_{32}) & -x_3 V_1 Y_{31} \sin(x_2 - \phi_{31}) & V_1 Y_{31} \cos(x_2 - \phi_{31}) \\ & -x_3 V_2 Y_{32} \sin(x_2 - x_1 - \phi_{32}) & +V_2 Y_{32} \cos(x_2 - x_1 - \phi_{32}) \\ & & +2x_3 Y_{33} \cos \phi_{33} \\ V_1 V_2 Y_{21} \cos(x_1 - \phi_{21}) & -V_2 x_3 Y_{23} \cos(x_1 - x_2 - \phi_{23}) & V_2 Y_{23} \sin(x_1 - x_2 - \phi_{23}) \\ +V_2 x_3 Y_{23} \cos(x_1 - x_2 - \phi_{23}) & & \end{bmatrix}$$

$$\tag{6.105}$$

This matrix has rank 3; therefore, this set of measurements spans the observable space of the power system.

The covariance matrix of the measurements is

$$R = \begin{bmatrix} \frac{1}{0.010^2} & & & & \\ & \frac{1}{0.050^2} & & & \\ & & \frac{1}{0.075^2} & & \\ & & & \frac{1}{0.050^2} & \\ & & & & \frac{1}{0.075^2} \end{bmatrix} \tag{6.106}$$

The Newton-Raphson iteration to solve for the set of states x that minimize the weighted errors is:

$$\left[H_x^T\left(x^k\right) R^{-1} H_x\left(x^k\right)\right]\left[x^{k-1} - x^k\right] = H_x^T\left(x^k\right) R^{-1}\left[z - h(x^k)\right] \tag{6.107}$$

Iteration 1

The initial condition for the state estimation solution is the same flat start as for the power flow equations; namely, all angles are set to zero and all unknown voltage magnitudes are set to unity. The measurement functions $h(x)$ evaluated at the initial conditions are

$$h(x^0) = \begin{bmatrix} 1.0000 \\ 0.0202 \\ -0.0664 \\ -0.0198 \\ -0.1914 \end{bmatrix}$$

The matrix of partials evaluated at the initial condition yields

$$H_x^0 = \begin{bmatrix} 0 & 0 & 1.0000 \\ 0 & -10.0990 & -1.0099 \\ -0.2257 & 0 & 0 \\ -9.9010 & 20.0000 & 1.9604 \\ -1.2158 & 0.9901 & -9.9010 \end{bmatrix}$$

The nonlinear functions (6.104) are

$$F(x^0) = \begin{bmatrix} 0.5655 \\ -1.4805 \\ -0.2250 \end{bmatrix}$$

The incremental updates for the states are

$$\Delta x^1 = \begin{bmatrix} -0.0119 \\ -0.0625 \\ -0.0154 \end{bmatrix}$$

leading to the updated states

$$\begin{bmatrix} \delta_2^1 \\ \delta_3^1 \\ V_3^1 \end{bmatrix} = \begin{bmatrix} -0.0119 \\ -0.0625 \\ 0.9846 \end{bmatrix}$$

where δ_2 and δ_3 are in radians. The error at iteration 0 is

$$\varepsilon^0 = 1.4805$$

Iteration 2

The updated values are used to recalculate the Newton-Raphson iterations:

$$h(x^1) = \begin{bmatrix} 0.9846 \\ 0.6585 \\ -0.0634 \\ -1.1599 \\ -0.0724 \end{bmatrix}$$

The matrix of partials is:

$$H_x^1 = \begin{bmatrix} 0 & 0 & 1.0000 \\ 0 & -9.9858 & -0.3774 \\ -0.2660 & 0 & 0 \\ -9.6864 & 19.5480 & 0.7715 \\ -0.7468 & 0.4809 & -9.9384 \end{bmatrix}$$

The nonlinear function evaluated at the updated values yields:

$$F(x^1) = \begin{bmatrix} 0.0113 \\ -0.0258 \\ 0.0091 \end{bmatrix}$$

The incremental updates for the states are

$$\Delta x^2 = \begin{bmatrix} 0.0007 \\ -0.0008 \\ 0.0013 \end{bmatrix}$$

leading to the updated states

$$\begin{bmatrix} \delta_2^2 \\ \delta_3^2 \\ V_3^2 \end{bmatrix} = \begin{bmatrix} -0.0113 \\ -0.0633 \\ 0.9858 \end{bmatrix}$$

The error at iteration 1 is

$$\varepsilon^1 = 0.0258$$

The iterations are obviously converging. At convergence, the states that minimize the weighted measurement errors are

$$x = \begin{bmatrix} -0.0113 \\ -0.0633 \\ 0.9858 \end{bmatrix}$$

To check for the presence of bad data, the weighted sum of squares of the measurement errors are compared to the chi-square distribution for $k = 2$ and $\alpha = 0.01$. The weighted sum of squares is

$$
\begin{aligned}
f &= \sum_{i=1}^{5} \frac{1}{\sigma_i^2} (z_i - h_i(x))^2 \\
&= \frac{(-0.0108)^2}{0.010^2} + \frac{(0.0015)^2}{0.050^2} + \frac{(-0.0184)^2}{0.075^2} + \frac{(0.0008)^2}{0.050^2} + \frac{(-0.0001)^2}{0.075^2} \\
&= 1.2335
\end{aligned}
$$

This value is less than $\chi_{2,0.01} = 9.21$; therefore, the data set is good and does not contain any spurious measurements. ∎

6.4   Problems

1. The fuel costs for a three-unit plant are given by

$$F_1 : 173.61 + 8.670P_1 + 0.00230P_1^2 \quad \$/MWhr$$
$$F_2 : 180.68 + 9.039P_2 + 0.00238P_2^2 \quad \$/MWhr$$
$$F_3 : 182.62 + 9.190P_3 + 0.00235P_3^2 \quad \$/MWhr$$

The daily load curve for the plant is given in Figure 6.6. Obtain and sketch the optimal power generated by each unit and the plant's incremental cost of power delivered (λ).

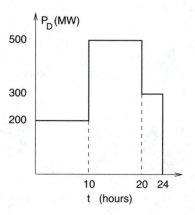

FIGURE 6.6
Load curve for Problem 1

2. Use the method of least squares to find the "best fit" coefficients c_0 and c_1 in the following function

$$f(x) = c_0 + c_1 x$$

for the following measured data:

x	$f(x)$
1	-2.1
3	-0.9
4	-0.6
6	0.6
7	0.9

3. Use the method of least squares to find the "best fit" coefficients a_0, a_1, and a_2 in the following function

$$f(t) = a_0 + a_1 \sin \frac{2\pi t}{12} + a_2 \cos \frac{2\pi t}{12}$$

for the following measured data:

t	$f(t)$
0	1.0
2	1.6
4	1.4
6	0.6
8	0.2
10	0.8

This function describes the movement of the tide with a 12-hour period.

4. Minimize $x_1^2 + x_2^2$ subject to the constraint

$$x_1^2 + 2x_1 x_2 + 3x_2^2 - 1 = 0$$

5. Find the minimum of

$$C: \quad x_1^2 + x_2^2 + u_1 x_1 + u_2 x_2 + 1$$

(a) subject to:

$$x_1 \cos(x_2) + x_2^2 - u_1 \cos(x_1) = 1$$
$$x_1 - x_2 + 3u_2 = -3$$

(b) subject to:

$$x_1 \cos(x_2) + x_2^2 - u_1 \cos(x_1) = 1$$
$$x_1 - x_2 + 3u_2 = -3$$
$$u_2 \geq -0.8$$

(c) subject to:

$$x_1 \cos(x_2) + x_2^2 - u_1 \cos(x_1) = 1$$
$$x_1 - x_2 + 3u_2 = -3$$
$$x_2 \le 0.30$$

using the penalty function $f(x_2) = ae^{b(x_2-c)}$, where a and b are positive constants and c is the function offset.

Use an initial guess of $x^0 = [0\ 0]'$ and $u^0 = [0\ 0]'$ and a $\gamma = 0.05$. You might want to experiment with other values of γ as well. Your stopping criterion should be $\|\nabla C\| \le 0.01$.

6. Consider the system shown in Figure 6.7. The bus and line data are given below:

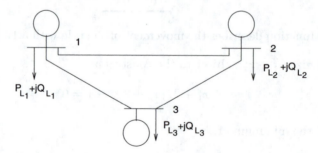

FIGURE 6.7
3 bus system

Line	R	X	B
1-2	0.01	0.1	0.050
1-3	0.05	0.1	0.025
2-3	0.05	0.1	0.025
Bus	$|V|$	P_L	Q_L
1	1.00	0.35	0.10
2	1.02	0.40	0.25
3	1.02	0.25	0.10

The fuel costs for the generators are:

$$F_1 : P_{G_1} + 1.5P_{G_1}^2$$
$$F_2 : 2P_{G_2} + P_{G_2}^2$$
$$F_3 : 2.5P_{G_3} + 0.5P_{G_3}^2$$

(a) Using the equal incremental cost criterion, find the optimal scheduling for the units (remember that this method neglects system losses).

(b) Using your answer for part (a) as the initial control vector, use the steepest descent method to find the optimal scheduling for this system, which considers system losses.

(c) Now assume the following limits are imposed:

$$F_1 : P_{G_1} + 1.5P_{G_1}^2 \quad 0 \le P_{G_1} \le 0.6$$
$$F_2 : 2P_{G_2} + P_{G_2}^2 \quad 0 \le P_{G_2} \le 0.4$$
$$F_3 : 2.5P_{G_3} + 0.5P_{G_3}^2 \quad 0 \le P_{G_3} \le 0.1$$

Repeat part (b).

(d) Interpret your results relating the generator settings to the Cost functions.

7. For the system shown in Figure 6.7, the following measurements were obtained:

V_2	1.04
V_3	0.98
P_{G1}	0.58
P_{G2}	0.30
P_{G3}	0.14
P_{12}	0.12
P_{32}	-0.04
P_{13}	0.10

where $\sigma_V^2 = (0.01)^2, \sigma_{P_G}^2 = (0.015)^2$, and $\sigma_{P_{ij}}^2 = (0.02)^2$.

Estimate the system states, the error, and test for bad data using the chi-square data below using $\alpha = 0.01$.

Chapter 7

Eigenvalue Problems

Small signal stability is the ability of a system to maintain stability when subjected to small disturbances. Small signal analysis provides valuable information about the inherent dynamic characteristics of the system and assists in its design, operation, and control. Time domain simulation and eigenanalysis are the two main approaches of studying system stability.

Eigenanalysis methods are widely used to perform small signal stability studies. The dynamic behavior of a system in response to small perturbations can be determined by computing the eigenvalues and eigenvectors of the system matrix. The locations of the eigenvalues can be used to investigate the system's performance. In addition, eigenvectors can be used to estimate the relative participation of the respective states in the corresponding disturbance modes.

A scalar λ is an eigenvalue of an $n \times n$ matrix A if there exists a nonzero $n \times 1$ vector v such that

$$Av = \lambda v \tag{7.1}$$

where v is the corresponding right eigenvector. If there exists a nonzero vector w such that

$$w^T A = \lambda w^T \tag{7.2}$$

then w is a left eigenvector. The set of all eigenvalues of A is called the spectrum of A. Normally the term "eigenvector" refers to the right eigenvector unless denoted otherwise. The eigenvalue problem in equation (7.1) is called the standard eigenvalue problem. Equation (7.1) can be written as

$$(A - \lambda I) x = 0 \tag{7.3}$$

and thus is a homogeneous system of equations for x. This system has a non-trivial solution only if the determinant

$$det\,(A - \lambda I) = 0$$

The determinant equation is also called the characteristic equation for A and is an n-th degree polynomial in λ. Therefore, there are n roots (possibly real or complex) of the characteristic equation. Each one of these roots is also an eigenvalue of A.

7.1   The QR Algorithm

The QR method [14], [39], [40] is one of the most widely used decomposition methods for calculating eigenvalues of matrices. It uses a sequence of orthogonal similarity transformations [8] [20]. Similar to the LU factorization, the matrix A can also be factored into two matrices such that

$$A = QR \tag{7.4}$$

where Q is a unitary matrix and R is an upper triangular matrix. The matrix Q is *unitary* if

$$QQ^* = Q^*Q = I \tag{7.5}$$

where $(*)$ denotes complex conjugate transpose.

Examples of unitary matrices are

$$Q_1 = \begin{bmatrix} 0 & 1 \\ 1 & 0 \end{bmatrix} \quad Q_2 = \begin{bmatrix} \cos\theta & -\sin\theta \\ \sin\theta & \cos\theta \end{bmatrix}$$

It also follows that the inverse of a unitary matrix is also its conjugate transpose, i.e.,

$$Q^{-1} = Q^*$$

This decomposition yields the column vectors $[a_1, a_2, \ldots, a_n]$ of A and column vectors $[q_1, q_2, \ldots, q_n]$ of Q such that

$$a_k = \sum_{i=1}^{k} r_{ik} q_i, \quad k = 1, \ldots, n \tag{7.6}$$

The column vectors $a_1, a_2, \ldots, a_n$ must be orthonormalized from the left to right into an orthonormal basis $q_1, q_2, \ldots, q_n$.

In the implementation of the QR algorithm, it is common practice to transform A into a Hessenberg matrix H having the same eigenvalues and then apply the QR matrix to H. In the end, the matrix becomes upper triangular and the eigenvalues can be read off of the diagonal. A Hessenberg matrix is essentially an upper triangular matrix with one extra set of non-zero elements directly below the diagonal. The reason for reducing A to a Hessenberg matrix is that this greatly reduces the total number of operations required for the QR algorithm.

The Householder method is one method used to reduce A to a Hessenberg matrix. For each $n \times n$ matrix A, there exist $n - 2$ Householder matrices $H_1, H_2, \ldots, H_{n-2}$, such that for

$$Q = H_{n-2} \ldots H_2 H_1$$

the matrix

$$P = Q^*AQ$$

is a Hessenberg matrix [19]. A matrix H is a Householder matrix if

$$H = I - 2\frac{vv^*}{v^*v}$$

Note that Householder matrices are also unitary matrices. The vector v is chosen to satisfy

$$v_i = a_i \pm e_i \, \|a_i\|_2 \tag{7.7}$$

where the choice of sign is based upon the requirement that $\|v\|_2$ should not be too small, e_i is the ith column of I, and a_i is the ith column of A.

Example 7.1
Find the QR decomposition of the matrix A:

$$A = \begin{bmatrix} 1 & 3 & 4 & 8 \\ 2 & 1 & 2 & 3 \\ 4 & 3 & 5 & 8 \\ 9 & 2 & 7 & 4 \end{bmatrix}$$

Solution 7.1 The first transformation will be applied to zero out the first column of A below the subdiagonal, thus

$$v_1 = a_1 + e_1 \, \|a_1\|_2$$

$$= \begin{bmatrix} 1 \\ 2 \\ 4 \\ 9 \end{bmatrix} + 10.0995 \begin{bmatrix} 1 \\ 0 \\ 0 \\ 0 \end{bmatrix}$$

$$= \begin{bmatrix} 11.0995 \\ 2.0000 \\ 4.0000 \\ 9.0000 \end{bmatrix}$$

leading to

$$H_1 = I - 2\frac{v_1 v_1^*}{(v_1^* v_1)}$$

$$= \begin{bmatrix} 1 & 0 & 0 & 0 \\ 0 & 1 & 0 & 0 \\ 0 & 0 & 1 & 0 \\ 0 & 0 & 0 & 1 \end{bmatrix} - \frac{2}{224.1990} \begin{bmatrix} 11.0995 \\ 2.0000 \\ 4.0000 \\ 9.0000 \end{bmatrix} \begin{bmatrix} 11.0995 & 2.0000 & 4.0000 & 9.0000 \end{bmatrix}$$

$$= \begin{bmatrix} -0.0990 & -0.1980 & -0.3961 & -0.8911 \\ -0.1980 & 0.9643 & -0.0714 & -0.1606 \\ -0.3961 & -0.0714 & 0.8573 & -0.3211 \\ -0.8911 & -0.1606 & -0.3211 & 0.2774 \end{bmatrix}$$

and

$$H_1 A = \begin{bmatrix} -10.0995 & -3.4655 & -9.0103 & -8.1192 \\ 0 & -0.1650 & -0.3443 & 0.0955 \\ 0 & 0.6700 & 0.3114 & 2.1910 \\ 0 & -3.2425 & -3.5494 & -9.0702 \end{bmatrix}$$

The second iteration will operate on the part of the transformed matrix that excludes the first column and row. Therefore

$$v_2 = a_2 + e_2 \, \|a_2\|_2$$

$$= \begin{bmatrix} -0.1650 \\ 0.6700 \\ -3.2425 \end{bmatrix} + 3.3151 \begin{bmatrix} 1 \\ 0 \\ 0 \end{bmatrix}$$

$$= \begin{bmatrix} 3.1501 \\ 0.6700 \\ -3.2425 \end{bmatrix}$$

which results in

$$H_2 = I - 2 \frac{v_2 v_2^*}{(v_2^* v_2)}$$

$$= \begin{bmatrix} 1 & 0 & 0 & 0 \\ 0 & 0.0498 & -0.2021 & 0.9781 \\ 0 & -0.2021 & 0.9570 & 0.2080 \\ 0 & 0.9781 & 0.2080 & -0.0068 \end{bmatrix}$$

and

$$H_2 H_1 A = \begin{bmatrix} -10.0995 & -3.4655 & -9.0103 & -8.1192 \\ 0 & -3.3151 & -3.5517 & -9.3096 \\ 0 & 0 & -0.3708 & 0.1907 \\ 0 & 0 & -0.2479 & 0.6108 \end{bmatrix}$$

Continuing the process yields

$$v_3 = \begin{bmatrix} 0.0752 \\ -0.2479 \end{bmatrix}$$

$$H_3 = \begin{bmatrix} 1 & 0 & 0 & 0 \\ 0 & 1 & 0 & 0 \\ 0 & 0 & 0.8313 & 0.5558 \\ 0 & 0 & 0.5558 & -0.8313 \end{bmatrix}$$

which results in

$$R = H_3 H_2 H_1 A = \begin{bmatrix} -10.0995 & -3.4655 & -9.0103 & -8.1192 \\ 0 & -3.3151 & -3.5517 & -9.3096 \\ 0 & 0 & -0.4460 & 0.4980 \\ 0 & 0 & 0 & -0.4018 \end{bmatrix}$$

and

$$Q = H_1 H_2 H_3 = \begin{bmatrix} -0.0990 & -0.8014 & -0.5860 & -0.0670 \\ -0.1980 & -0.0946 & 0.2700 & -0.9375 \\ -0.3961 & -0.4909 & 0.7000 & 0.3348 \\ -0.8911 & 0.3283 & -0.3060 & 0.0670 \end{bmatrix}$$

It can be verified that $A = QR$ and further that $Q^* = Q^{-1}$. ∎

The elimination by QR decomposition can be considered as an alternative to Gaussian elimination. However, the number of multiplications and divisions required is more than twice the number required for Gaussian elimination. Therefore QR decomposition is seldom used for the solution of linear systems, but it does play an important role in the calculation of eigenvalues.

Although the eigenvalue problem gives rise to a simple set of algebraic equations to determine the solution to

$$det\,(A - \lambda I) = 0$$

the practical problem of solving this equation is difficult. Computing the roots of the characteristic equation or the nullspace of a matrix is a process that is not well suited for computers. In fact, no generalized direct process exists for solving the eigenvalue problem in a finite number of steps. Therefore iterative methods for calculation must be relied upon to produce a series of successively improved approximations to the eigenvalues of a matrix.

The QR method is commonly used to calculate the eigenvalues and eigenvectors of full matrices. As developed by Francis [8], the QR method produces a series of similarity transformations

$$A_k = Q_k^* A_{k-1} Q_k \quad Q_k^* Q_k = I \tag{7.8}$$

where the matrix A_k is similar to A. The QR decomposition is repeatedly performed and applied to A as the subdiagonal elements are iteratively driven to zero. At convergence, the eigenvalues of A in descending order by magnitude appear on the diagonal of A_k.

Example 7.2
Find the eigenvalues and eigenvectors of the matrix of Example 7.1.

Solution 7.2 The first objective is to find the eigenvalues of the matrix A using the QR method. From Example 7.1, the first QR factorization yields the Q_0 matrix

$$Q_0 = \begin{bmatrix} -0.0990 & 0.8014 & 0.5860 & -0.0670 \\ -0.1980 & 0.0946 & -0.2700 & -0.9375 \\ -0.3961 & 0.4909 & -0.7000 & 0.3348 \\ -0.8911 & -0.3283 & 0.3060 & 0.0670 \end{bmatrix}$$

Using the given A matrix as A_0, the first update A_1 is found by

$$A_1 = Q_0^* A_0 Q_0$$
$$= \begin{bmatrix} 12.4902 & -10.1801 & -1.1599 & 0.3647 \\ -10.3593 & -0.9987 & -0.5326 & -1.2954 \\ 0.2672 & 0.3824 & -0.4646 & 0.1160 \\ 0.3580 & 0.1319 & -0.1230 & -0.0269 \end{bmatrix}$$

The QR factorization of A_1 yields

$$Q_1 = \begin{bmatrix} -0.7694 & -0.6379 & -0.0324 & -0.0006 \\ 0.6382 & -0.7660 & -0.0733 & 0.0252 \\ -0.0165 & 0.0687 & -0.9570 & -0.2812 \\ -0.0221 & 0.0398 & -0.2786 & 0.9593 \end{bmatrix}$$

and the A_2 matrix becomes

$$A_2 = Q_1^* A_1 Q_1$$
$$= \begin{bmatrix} 17.0913 & 4.8455 & -0.2315 & -1.0310 \\ 4.6173 & -5.4778 & -1.8116 & 0.6064 \\ -0.0087 & 0.0373 & -0.5260 & -0.1757 \\ 0.0020 & -0.0036 & 0.0254 & -0.0875 \end{bmatrix}$$

Note that the elements below the diagonals are slowly decreasing to zero. This process is carried out until the final A matrix is obtained:

$$A_* = \begin{bmatrix} 18.0425 & 0.2133 & -0.5180 & -0.9293 \\ 0 & -6.4172 & -1.8164 & 0.6903 \\ 0 & 0 & -0.5269 & -0.1972 \\ 0 & 0 & 0 & -0.0983 \end{bmatrix} \tag{7.9}$$

The eigenvalues are on the diagonals of A_* and are in decreasing order by magnitude. Thus the eigenvalues are

$$\lambda_{1,\dots,4} = \begin{bmatrix} 18.0425 \\ -6.4172 \\ -0.5269 \\ -0.0983 \end{bmatrix}$$

The next step is to find the eigenvectors associated with each eigenvalue. Recall that

$$A v_i = \lambda_i v_i \tag{7.10}$$

for each eigenvalue and corresponding eigenvector $i = 1, \ldots, n$. Equation (7.10) may also be written as

$$Av_i - \lambda_i v_i = 0$$

In other words, the matrix defined by $A - \lambda_i I$ is singular; thus, only three of its rows (or columns) are independent. This fact can be used to determine the eigenvectors once the eigenvalues are known. Since $A - \lambda_i I$ is not of full rank, one of the elements of the eigenvector v_i can be chosen arbitrarily. To start, partition $A - \lambda_i I$ as

$$A - \lambda_i I = \begin{bmatrix} a_{11} & a_{1,2n} \\ a_{2n,1} & a_{2n,2n} \end{bmatrix}$$

where a_{11} is a scalar, $a_{1,2n}$ is a $1 \times (n-1)$ vector, $a_{2n,1}$ is a $(n-1) \times 1$ vector, and $a_{2n,2n}$ is an $(n-1) \times (n-1)$ matrix of rank $(n-1)$. Then let $v_i(1) = 1$ and solve for the remaining portion of the eigenvector as

$$\begin{bmatrix} v_i(2) \\ v_i(3) \\ \vdots \\ v_i(n) \end{bmatrix} = -a_{2n,2n}^{-1} a_{2n,1} v_i(1) \tag{7.11}$$

Now update $v_i(1)$ from

$$v_i(1) = -\frac{1}{a_{11}} a_{2n,1} * \begin{bmatrix} v_i(2) \\ v_i(3) \\ \vdots \\ v_i(n) \end{bmatrix}$$

Then the eigenvector corresponding to λ_i is

$$v_i = \begin{bmatrix} v_i(1) \\ v_i(2) \\ \vdots \\ v_i(n) \end{bmatrix}$$

The last step is to normalize the eigenvector; therefore,

$$v_i = \frac{v_i}{\|v_i\|}$$

Thus, for the vector of eigenvalues

$$\Lambda = \begin{bmatrix} 18.0425 & -6.4172 & -0.5269 & -0.0983 \end{bmatrix}$$

the corresponding eigenvectors are:

$$\begin{bmatrix} 0.4698 \\ 0.2329 \\ 0.5800 \\ 0.6234 \end{bmatrix}, \begin{bmatrix} 0.6158 \\ 0.0539 \\ 0.2837 \\ -0.7330 \end{bmatrix}, \begin{bmatrix} 0.3673 \\ -0.5644 \\ -0.5949 \\ 0.4390 \end{bmatrix}, \begin{bmatrix} 0.0932 \\ 0.9344 \\ -0.2463 \\ -0.2400 \end{bmatrix} \blacksquare$$

7.1.1   Shifted QR

The QR iterations can converge very slowly in many instances. However, if some information about one or more of the eigenvalues is known a priori, then a variety of techniques can be applied to speed up convergence of the iterations. One such technique is the *shifted* QR method, in which a shift σ is introduced at each iteration such that the QR factorization at the kth is performed on

$$A_k - \sigma I = Q_k R_k$$

and

$$A_{k+1} = Q_k^* (A_k - \sigma I) Q_k + \sigma I$$

If σ is a good estimate of an eigenvalue, then the $(n, n-1)$ entry of A_k will converge rapidly to zero, and the (n, n) entry of A_k will converge to the eigenvalue closest to σ_k. Once this has occurred, an alternate shift can be applied.

Example 7.3
Repeat Example 7.2 using shifts.

Solution 7.3 Start with using a shift of $\sigma = 15$. This is near the 18.0425 eigenvalue; so, convergence to that particular eigenvalue should be rapid. Starting with the original A matrix as A_0, the QR factorization of $A_0 - \sigma I$ yields

$$Q_0 = \begin{bmatrix} -0.8124 & 0.0764 & 0.2230 & 0.5334 \\ 0.1161 & -0.9417 & -0.0098 & 0.3158 \\ 0.2321 & 0.2427 & -0.7122 & 0.6164 \\ 0.5222 & 0.2203 & 0.6655 & 0.4856 \end{bmatrix}$$

and the update $A_1 = Q_0^* (A_0 - \sigma I) Q_0 + \sigma I$

$$A_1 = \begin{bmatrix} -4.9024 & 0.8831 & -1.6174 & 2.5476 \\ -0.2869 & 0.0780 & -0.1823 & 1.7775 \\ -2.9457 & 0.5894 & -1.5086 & 2.3300 \\ 2.5090 & 1.0584 & 3.1975 & 17.3330 \end{bmatrix}$$

The eigenvalue of interest ($\lambda = 18.00425$) will now appear in the lower right corner since as the iterations progress $A_{k+1}(n, n) - \sigma$ will be the smallest diagonal in magnitude. Recall that the eigenvalues are ordered on the diagonal from largest to smallest and since the largest eigenvalue is "shifted" by σ, it will now have the smallest magnitude. The convergence can be further increased by updating σ at each iteration, such that $\sigma_{k+1} = A_{k+1}(n, n)$. The iterations proceed as in Example 7.2. ∎

7.1.2 Deflation

The speed of convergence of the QR method for calculating eigenvalues depends greatly on the location of the eigenvalues with respect to one another. The matrix $A - \sigma I$ has the eigenvalues $\lambda_i - \sigma$ for $i = 1, \ldots, n$. If σ is chosen as an approximate value of the smallest eigenvalue λ_n, then $\lambda_n - \sigma$ becomes small. This will speed up the convergence in the last row of the matrix, since

$$\frac{|\lambda_n - \sigma|}{|\lambda_{n-1} - \sigma|} \ll 1$$

Once the elements of the last row are reduced to zero, the last row and column of the matrix may be neglected. This implies that the smallest eigenvalue is "deflated" by removing the last row and column. The procedure can then be repeated on the remaining $(n-1) \times (n-1)$ matrix with the shift σ chosen close to λ_{n-1}. Using the shift and deflation in combination can significantly improve convergence. Additionally, if only one eigenvalue is desired of a particular magnitude, this eigenvalue can be isolated via the shift method. After the last row has been driven to zero, the eigenvalue can be obtained and the remainder of the QR iterations abandoned.

7.2 Arnoldi Methods

In large interconnected systems, it is either impractical or intractable to find all of the eigenvalues of the system state matrix due to restrictions on computer memory and computational speed. The Arnoldi method has been developed as an algorithm that iteratively computes k eigenvalues of an $n \times n$ matrix A, where k is typically much smaller than n. This method therefore bypasses many of the constraints imposed by large matrix manipulation required by methods such as the QR decomposition. If the k eigenvalues are chosen selectively, they can yield rich information about the system under consideration, even without the full set of eigenvalues. The Arnoldi method was first developed in [2], but suffered from poor numerical properties such as loss of orthogonality and slow convergence. Several modifications to the Arnoldi method have overcome these shortcomings. The Modified Arnoldi Method (MAM) has been used frequently in solving eigenvalue problems in power system applications [21], [38]. This approach introduced preconditioning and explicit restart techniques to retain orthogonality. Unfortunately however, an explicit restart will often discard useful information. The restart problem was solved by using implicitly shifted QR steps [33] in the Implicitly Restarted Arnoldi (IRA) method. Several commercial software packages have been developed around the IRA method, including the well known ARPACK and the Matlab `speig` routines.

The basic approach of the Arnoldi method is to iteratively update a low order matrix H whose eigenvalues successively approximate the selected eigenvalues of the larger A matrix, such that

$$AV = VH; \quad V^*V = I \tag{7.12}$$

where V is an $n \times k$ matrix and H is a $k \times k$ Hessenberg matrix. As the method progresses, the eigenvalues of A are approximated by the diagonal entries of H yielding

$$HV_i = V_i D \tag{7.13}$$

where V_i is a $k \times k$ matrix whose columns are the eigenvalues of H (approximating the eigenvectors of A) and D is a $k \times k$ matrix whose diagonal entries are the eigenvalues of H (approximating the eigenvalues of A). The Arnoldi method is an orthogonal projection method onto a *Krylov* subspace.

The Arnoldi procedure is an algorithm for building an orthogonal basis of the Krylov subspace. One approach is given as:

The k-step Arnoldi Factorization

Starting with a vector v_1 of unity norm, for $j = 1, \ldots, k$ compute:

1. $H(i,j) = v_i^T A v_j$ for $i = 1, \ldots, j$

2. $w_j = A v_j - \sum_{i=1}^{j} H(i,j) v_i$

3. $H(j+1,j) = \|w_j\|_2$

4. If $H(j+1,j) = 0$, then stop

5. $v_{j+1} = \frac{w_j}{H(j+1,j)}$

At each step, the algorithm multiplies the previous Arnoldi vector v_j by A and then orthonormalizes the resulting vector w_j against all previous v_i's. The k-step Arnoldi factorization is shown in Figure 7.1, and is given by

$$AV_k = V_k H_k + w_k e_k^T \tag{7.14}$$

The columns $V = [v_1, v_2, \ldots, v_k]$ form an orthonormal basis for the Krylov subspace and H is the orthogonal projection of A onto this space. It is desirable for $\|w_k\|$ to become small because this indicates that the eigenvalues of H are accurate approximations to the eigenvalues of A. However, this "convergence" often comes at the price of numerical orthogonality in V. Therefore, the k-step Arnoldi factorization is "restarted" to preserve orthogonality.

Implicit restarting provides a means to extract rich information from very large Krylov subspaces while avoiding the storage and poor numerical properties associated with the standard approach. This is accomplished by continually compressing the information into a fixed size k-dimensional subspace, by using a shifted QR mechanism. A $(k + p)$-step Arnoldi factorization

$$AV_{k+p} = V_{k+p} H_{k+p} + w_{k+p} e_{k+p}^T \tag{7.15}$$

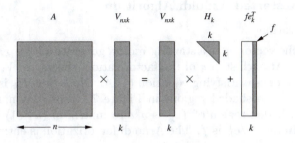

FIGURE 7.1

A k-step Arnoldi factorization

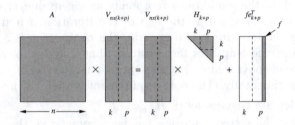

FIGURE 7.2

A $(k + p)$-step Arnoldi factorization

is compressed to a factorization of length k that retains the eigen-information of interest. This is accomplished using QR steps to apply p shifts to yield

$$A\hat{V}_{k+p} = \hat{V}_{k+p}\hat{H}_{k+p} + \hat{w}_{k+p} \qquad (7.16)$$

where $\hat{V}_{k+p} = V_{k+p}Q$, $\hat{H}_{k+p} = Q^*H_{k+p}Q$, and $\hat{w}_{k+p} = w_{k+p}e_{k+p}^TQ$. It may be shown that the first $k - 1$ entries of the vector e_{k+p}^TQ are zero [34]. Equating the first k columns on both sides yields an updated k-step Arnoldi factorization. This now provides the "restart" vectors for extending the k-step Arnoldi factorization to the $k + p$-step Arnoldi factorization, shown in Figure 7.2.

The implicitly restarted Arnoldi algorithm consists of three main steps: initialization, iteration/refinement, and final calculation of the eigenvalues and eigenvectors.

Implicity Restarted Arnoldi Algorithm

1. Initialization

   Using the vector v_1 as a starting vector, generate a k-step Arnoldi factorization. At each step k of the factorization, the vector V_k is augmented by a vector v_k satisfying equation (7.14). Note that H_k is a Hessenberg matrix. The shaded regions in Figure 7.1 represent non-zero entries. The unshaded region of fe_k^T is a zero matrix of $(k-1)$ columns. The last column of fe_k^T is f. The Arnoldi factorization is entirely dependent on the choice of initial vector v_1.

2. Iteration/Refinement

   (a) Extend the k-step factorization by p steps.

   Each of the p additions represents an eigenvalue/eigenvector that can be discarded at the end of the iteration if it does not meet the chosen criteria. In general, the choice of p is a trade-off between the length of factorization that may be tolerated and the rate of convergence. For most problems, the size of p is determined experimentally. The only requirement is that $1 \leq p \leq n - k$.

   (b) Calculate eigenvalues of H_{k+p}

   After the p-step extension has been completed, the eigenvalues of H_{k+p} are calculated by the QR method and sorted according to a pre-determined sort criterion S and ordered from best to worst. The p worst eigenvalues $(\sigma_1, \sigma_2, \ldots, \sigma_p)$ are used as shifts to perform p shifted QR factorizations. Since the matrix H_{k+p} in the Arnoldi factorization

   $$AV_{k+p} = V_{k+p}H_{k+p} + w_{k+p}e_{k+p}^T \tag{7.17}$$

   is relatively small, the shifted QR factorization can be used efficiently to calculate the eigenvalues of H.

   (c) Update the Arnoldi matrices

   $$\hat{V}_{k+p} = V_{k+p}Q$$
   $$\hat{H}_{k+p} = Q^*H_{k+p}Q$$
   $$\hat{w}_{k+p} = w_{k+p}e_{k+p}^TQ$$

   Note that the updated matrix $\hat{V}_{k+p}$ has orthonormal columns since it is the product of V and an orthogonal matrix Q.

   (d) Obtain a new k-step Arnoldi factorization by equating the first k columns on each side of equation (7.16) and discarding the last p equations:

   $$A\hat{V}_k = \hat{V}_k\hat{H}_k + \hat{w}_k e_k^T$$

   The vector $\hat{w}$ is the new residual vector that is being driven to zero.

(e) If

$$\|AV_k - V_k H_k\| \le \varepsilon$$

where ε is the pre-selected convergence tolerance, then the iteration/refinement terminates. Otherwise the process is repeated until tolerance is achieved.

3. Eigenvalue/Eigenvector Calculation

The last step in the Arnoldi method is to compute the eigenvalues and eigenvectors of the reduced matrix H_k from

$$H_k V_k + V_h D_k \qquad (7.18)$$

The eigenvectors of A are then calculated as

$$V_k = V_k V_h \qquad (7.19)$$

and the desired eigenvalues of A may be obtained from the diagonal entries of D_k:

$$AV_k = V_k D_k \qquad (7.20)$$

Example 7.4
Using a 3-step Arnoldi factorization, find the two smallest (in magnitude) eigenvalues and corresponding eigenvectors of the matrix of Example 7.1.

Solution 7.4 Since the two smallest eigenvalues are desired, the value of k is two. After the initialization step, the 2-step Arnoldi method will be extended up to 3-steps; therefore, p is one. Thus at each step, three eigenvalues will be calculated and the worst eigenvalue will be discarded.

The factorization can be initialized with an arbitrary non-zero vector. In many software implementations, the starting vector is chosen randomly such that all of the entries have absolute value less than 0.5. The starting vector for this example will be

$$v_0 = \begin{bmatrix} 0.2500 \\ 0.2500 \\ 0.2500 \\ 0.2500 \end{bmatrix}$$

To satisfy the requirement that the initial vector have unity norm, the starting vector is normalized to yield:

$$v_1 = \frac{Av_0}{\|Av_0\|}$$

$$= \begin{bmatrix} 0.4611 \\ 0.2306 \\ 0.5764 \\ 0.6340 \end{bmatrix}$$

After the initial vector has been chosen, the Arnoldi factorization is applied for k steps; thus,

$$h_{2,1}v_2 = Av_1 - h_{1,1}v_1 \qquad (7.21)$$

where v_2 produces the second column of the matrix V_k and $h_{1,1}$ is chosen such that

$$h_{1,1} = \langle v_1, Av_1 \rangle = v_1^T Av_1 \qquad (7.22)$$

where $\langle \cdot \rangle$ denotes inner product. Thus, solving equation (7.22) yields $h_{1,1} = 18.0399$. Applying the Arnoldi factorization for f_1 yields

$$w_1 = h_{2,1}v_2 = Av_1 - h_{1,1}v_1$$

$$= \begin{bmatrix} 1 & 3 & 4 & 8 \\ 2 & 1 & 2 & 3 \\ 4 & 3 & 5 & 8 \\ 9 & 2 & 7 & 4 \end{bmatrix} \begin{bmatrix} 0.4611 \\ 0.2306 \\ 0.5764 \\ 0.6340 \end{bmatrix} - (18.0399) \begin{bmatrix} 0.4611 \\ 0.2306 \\ 0.5764 \\ 0.6340 \end{bmatrix}$$

$$= \begin{bmatrix} 0.2122 \\ 0.0484 \\ 0.0923 \\ -0.2558 \end{bmatrix}$$

The factor $h_{2,1}$ is chosen to normalize v_2 to unity, thus $h_{2,1} = 0.3483$, and

$$v_2 = \begin{bmatrix} 0.6091 \\ 0.1391 \\ 0.2650 \\ -0.7345 \end{bmatrix}$$

Calculating the remaining values of the Hessenberg matrix yields:

$$h_{1,2} = v_1^* Av_2 = 0.1671$$
$$h_{2,2} = v_2^* Av_2 = -6.2370$$

and

$$w_2 = h_{3,2}v_2 = Av_2 - h_{1,2}v_1 - h_{2,2}v_2 = \begin{bmatrix} -0.0674 \\ 0.5128 \\ -0.1407 \\ -0.0095 \end{bmatrix}$$

These values can be checked to verify that they satisfy equation (7.14) for $i = 2$:

$$AV_2 = V_2 H_2 + w_2 \begin{bmatrix} 0 & 1 \end{bmatrix}$$

where

$$V_2 = \begin{bmatrix} v_1 & v_2 \end{bmatrix} = \begin{bmatrix} 0.4611 & 0.6091 \\ 0.2306 & 0.1391 \\ 0.5764 & 0.2650 \\ 0.6340 & -0.7345 \end{bmatrix}$$

and

$$H_2 = \begin{bmatrix} h_{1,1} & h_{1,2} \\ h_{2,1} & h_{2,2} \end{bmatrix} = \begin{bmatrix} 18.0399 & 0.1671 \\ 0.3483 & -6.2370 \end{bmatrix}$$

This completes the initialization stage.

After the initial k-step Arnoldi sequence has been generated, it can be extended to $k + p$ steps. In this example $p = 1$, so only one more extension is required. From the initialization, $w_2 = h_{3,2}v_2$ from which $h_{3,2}$ and v_2 can be extracted (recalling that $\|v_2\| = 1.0$) to yield $h_{3,2} = 0.5361$ and

$$v_3 = \begin{bmatrix} -0.1257 \\ 0.9565 \\ -0.2625 \\ -0.0178 \end{bmatrix}$$

The Hessenberg matrix H_3 becomes

$$H_3 = \begin{bmatrix} h_{1,1} & h_{1,2} & h_{1,3} \\ h_{2,1} & h_{2,2} & h_{2,3} \\ 0 & h_{3,2} & h_{3,3} \end{bmatrix} = \begin{bmatrix} 18.0399 & 0.1671 & 0.5560 \\ 0.3483 & -6.2370 & 2.0320 \\ 0 & 0.5361 & -0.2931 \end{bmatrix}$$

where

$$h_{1,3} = v_1^T A v_3$$
$$h_{2,3} = v_2^T A v_3$$
$$h_{3,3} = v_3^T A v_3$$

and

$$w_3 = A v_3 - h_{1,3}v_1 - h_{2,3}v_2 - h_{3,3}v_3 = \begin{bmatrix} 0.0207 \\ -0.0037 \\ -0.0238 \\ 0.0079 \end{bmatrix}$$

The next step is to compute (using QR factorization) and sort the eigenvalues (and eigenvectors) of the small matrix H_3. The eigenvalues of H_3 are:

$$\sigma = \begin{bmatrix} 18.0425 \\ -6.4166 \\ -0.1161 \end{bmatrix}$$

Since the smallest two eigenvalues are desired, the eigenvalues are sorted such that the desired eigenvalues are at the bottom (which they already are). The undesired eigenvalue estimate is $\sigma_1 = 18.0425$. Applying the shifted QR factorization to $H_3 - \sigma_1 I$ yields:

$$H_3 - \sigma_1 I = \begin{bmatrix} -0.0026 & 0.1671 & 0.5560 \\ 0.3483 & -24.2795 & 2.0320 \\ 0 & 0.5361 & -18.3356 \end{bmatrix}$$

and

$$QR = \begin{bmatrix} -0.0076 & -0.0311 & 0.9995 \\ 1.0000 & -0.0002 & 0.0076 \\ 0 & 0.9995 & 0.0311 \end{bmatrix} \begin{bmatrix} 0.3483 & -24.2801 & 2.0277 \\ 0 & 0.5364 & -18.3445 \\ 0 & 0 & 0 \end{bmatrix}$$

From Q, the update $\hat{H}$ can be found:

$$\hat{H}_3 = Q^* H_3 Q$$
$$= \begin{bmatrix} -6.2395 & 2.0216 & 0.2276 \\ 0.5264 & -0.2932 & -0.5673 \\ 0 & 0 & 18.0425 \end{bmatrix}$$

Note that the $\hat{H}(3,2)$ element is now zero. Continuing the algorithm yields the update for $\hat{V}$:

$$\hat{V} = V_3 Q = \begin{bmatrix} 0.6056 & -0.1401 & 0.4616 \\ 0.1373 & 0.9489 & 0.2613 \\ 0.2606 & -0.2804 & 0.5699 \\ -0.7392 & -0.0374 & 0.6276 \end{bmatrix}$$

where $V_3 = [v_1 \ v_2 \ v_3]$, and

$$\hat{w}_3 e^T = w_3 e_3^T Q = \begin{bmatrix} 0 & 0.0207 & 0.0006 \\ 0 & -0.0037 & -0.0001 \\ 0 & -0.0238 & -0.0007 \\ 0 & 0.0079 & 0.0002 \end{bmatrix}$$

Note that the first column of $\hat{w}e^T$ is zeros, so that a new k-step Arnoldi factorization can be obtained by equating the first k columns on each side such that

$$A\hat{V}_2 = \hat{V}_2 \hat{H}_2 + \hat{f}_2 e_2^T \tag{7.23}$$

The third columns of $\hat{V}$ and $\hat{H}$ are discarded.

This iteration/refinement procedure is continued until

$$\|AV - VH\| = \|we^T\| < \varepsilon$$

at which time the calculated eigenvalues will be obtained within order of ε accuracy. ∎

7.3   Linear Model Identification

Eigenvalues are calculated from linear matrices that arise from the dynamic model of a system. The system itself may be linear or the system may be linearized around an operating point of interest. In these cases, the system model is developed directly from the physical properties that govern the system behavior. In practice, however, the system may be too complex to model or may have parameters that drift with time or operating condition. In these cases, it may be desirable to replace the actual dynamic model with an estimated linear model that is derived from the system output waveform. The estimated linear model may then be used for control design applications or other linear analysis techniques. The estimated model may be chosen to be of lower order than the original model, but still retain the dominant modal characteristics.

This problem may be posed such that given a set of measurements that vary with time, is it possible to fit a time-varying waveform to the actual waveform (i.e., minimize the error between the actual measured waveform and the proposed waveform)? This process then results in a time-varying expression for a particular system state of the linear time invariant system

$$\dot{x}(t) = Ax(t) \quad x(t_0) = x_0 \tag{7.24}$$

where

$$x(t) = \sum_{i=1}^{n} \left(q_i^T x_0 \right) p_i e^{\lambda_i t} \tag{7.25}$$

$$= \sum_{i=1}^{n} R_i e^{\lambda_i t} \tag{7.26}$$

$$\tag{7.27}$$

is one of the n states, q_i, p_i, are the left and right eigenvectors corresponding to the eigenvalue λ respectively, and R is an $n \times n$ residue matrix. The estimation of these responses yields modal information about the system that can be used to predict possible unstable behavior, controller design, parametric summaries for damping studies, and modal interaction information.

Any time-varying function can be fit to a series of complex exponential functions over a finite time interval. However, it is not practical to include a large number of terms in the fitting function. The problem then becomes one of minimizing the error between the actual time-varying function and the proposed function by estimating the magnitude, phase, and damping parameters of the fitting function.

7.3.1 Prony Method

One approach to estimate the various parameters is the *Prony method* [16]. This method is designed to directly estimate the parameters for the exponential terms by fitting the function

$$\hat{y}(t) = \sum_{i=1}^{n} A_i e^{\sigma_i t} \cos \left(\omega_i t + \phi_i \right) \tag{7.28}$$

to an observed measurement for $y(t)$, where $y(t)$ consists of N samples

$$y(t_k) = y(k), \quad k = 0, 1, \ldots, N - 1$$

that are evenly spaced by a time interval Δt. Since the measurement signal $y(t)$ may contain noise or dc offset, it may have to be conditioned before the fitting process in applied.

The basic Prony method is summarized as

Prony Method

1. Construct a discrete linear prediction model from the measurement set

2. Find the roots of the characteristic polynomial of the model

3. Using the roots as the complex modal frequencies for the signal, determine the amplitude and phase for each mode

These steps are performed in the z-domain, translating the eigenvalues to the s-domain as a final step.

Note that equation (7.28) can be recast in complex exponential form as:

$$\hat{y}(t) = \sum_{i=1}^{n} B_i e^{\lambda_i t} \tag{7.29}$$

which can be translated to

$$\hat{y}(k) = \sum_{i=1}^{n} B_i z_i^k \tag{7.30}$$

where

$$z_i = e^{(\lambda_i \Delta t)} \tag{7.31}$$

The system eigenvalues λ can be found from the discrete modes by

$$\lambda_i = \frac{ln(z_i)}{\Delta t} \tag{7.32}$$

The z_i are the roots of the n-th order polynomial

$$z^n - \left(a_1 z^{n-1} + a_2 z^{n-2} + \ldots + a_n z^0 \right) = 0 \tag{7.33}$$

where the a_i coefficients are unknown and must be calculated from the measurement vector as:

$$\begin{bmatrix} y(n-1) & y(n-2) & \ldots & y(0) \\ y(n-0) & y(n-1) & \ldots & y(1) \\ \vdots & \vdots & \vdots & \vdots \\ y(N-2) & y(N-3) & \ldots & y(N-n-1) \end{bmatrix} \begin{bmatrix} a_1 \\ a_2 \\ \vdots \\ a_n \end{bmatrix} = \begin{bmatrix} y(n+0) \\ y(n+1) \\ \ldots \\ y(N-1) \end{bmatrix} \tag{7.34}$$

Note that this is a system of N equations in n unknowns and therefore must be solved by the least squares method to find the best fit.

Once z_i has been computed from the roots of equation (7.33), then the eigenvalues λ_i can be calculated from equation (7.32). The next step is to find the B_i that produces $\hat{y}(k) = y(k)$ for all k. This leads to the following relationship:

$$
\begin{bmatrix}
z_1^0 & z_2^0 & \cdots & z_n^0 \\
z_1^1 & z_2^1 & \cdots & z_n^1 \\
\vdots & \vdots & \vdots & \vdots \\
z_1^{N-1} & z_2^{N-1} & \cdots & z_n^{N-1}
\end{bmatrix}
\begin{bmatrix}
B_1 \\
B_2 \\
\vdots \\
B_n
\end{bmatrix}
=
\begin{bmatrix}
y(0) \\
y(1) \\
\vdots \\
y(N-1)
\end{bmatrix}
\tag{7.35}
$$

which can be succinctly expressed as

$$ZB = Y \tag{7.36}$$

Note that the matrix B is $n \times N$; therefore, equation (7.36) must also be solved by the least squares method. The estimating waveform $\hat{y}(t)$ is then calculated from equation (7.29). The reconstructed signal $\hat{y}(t)$ will usually not fit $y(t)$ exactly. An appropriate measure for the quality of this fit is a "signal to noise ratio (SNR)" given by

$$\text{SNR} = 20 \log \frac{\|\hat{y} - y\|}{\|y\|} \tag{7.37}$$

where the SNR is given in decibels (dB).

Since the fit for this method may be inexact, it is desirable to have control over the level of error between the fitting function and the original waveform. In this case, a nonlinear least squares can provide improved results.

7.3.2 The Levenberg-Marquardt Method

The nonlinear least squares for data fitting applications has the general form

$$\text{minimize } f(x) = \sum_{k=1}^{N} [\hat{y}(x, t_i) - y_i]^2 \tag{7.38}$$

where y_i is the output of the system at time t_i, and x is the vector of magnitudes, phases, and damping coefficients of equation (7.28), which arise from the eigenvalues of the state matrix of the system.

To find the minimum of $f(x)$, the same procedure for developing the Newton-Raphson iteration is applied. The function $f(x)$ is expanded about some x_0 by the Taylor series:

$$f(x) \approx f(x_0) + (x - x_0)^T f'(x_0) + \frac{1}{2}(x - x_0)^T f''(x_0)(x - x_0) + \cdots \tag{7.39}$$

where

$$f'(x) = \frac{\partial f}{\partial x_j} \quad \text{for } j = 1, \ldots, n$$

$$f''(x) = \frac{\partial^2 f}{\partial x_j \partial x_k} \quad \text{for } j, k = 1, \ldots, n$$

If the higher order terms in the Taylor's expansion are neglected, then minimizing the quadratic function on the right hand side of equation (7.39) yields

$$x_1 = x_0 - [f''(x_0)]^{-1} f'(x_0) \tag{7.40}$$

which yields an approximation for the minimum of the function $f(x)$. This is also one Newton-Raphson iteration update for solving the necessary minimization condition

$$f'(x) = 0$$

The Newton-Raphson equation (7.40) may be rewritten as the iterative linear system

$$A(x_k)(x_{k+1} - x_k) = g(x_k) \tag{7.41}$$

where

$$g_j(x) = -\frac{\partial f}{\partial x_j}(x)$$

$$a_{jk}(x) = \frac{\partial^2 f}{\partial x_j \partial x_k}(x)$$

and the matrix A is the system Jacobian (or similar iterative matrix).

The derivatives of equation (7.38) are

$$\frac{\partial f}{\partial x_j}(x) = 2 \sum_{k=1}^{N} [\hat{y}_k - y_k] \frac{\partial \hat{y}_i}{\partial x_j}(x)$$

and

$$\frac{\partial^2 f}{\partial x_j \partial x_k}(x) = 2 \sum_{k=1}^{N} \left\{ \frac{\partial \hat{y}_k}{\partial x_j}(x) \frac{\partial \hat{y}_i}{\partial x_k}(x) + [\hat{y}_k - y_k] \frac{\partial^2 \hat{y}_k}{\partial x_j \partial x_k}(x) \right\}$$

In this case, the matrix element a_{jk} contains second derivatives of the functions $\hat{y}_i$. These derivatives are multiplied by the factor $[\hat{y}_i(x) - y_i]$ and will become small during the minimization of f. Therefore the argument can be made that these terms can be neglected during the minimization process. Note that if the method converges, it will converge regardless of whether the exact Jacobian is used in the iteration. Therefore the iterative matrix A can be simplified as

$$a_{jk} = 2 \sum_{i=1}^{N} \frac{\partial \hat{y}_i}{\partial x_j}(x) \frac{\partial \hat{y}_i}{\partial x_k}(x) \tag{7.42}$$

and note that $a_{jj}(x) > 0$.

The Levenberg-Marquardt method modifies equation (7.41) by introducing the matrix $\hat{A}$ with entries

$$\hat{a}_{jj} = (1 + \gamma)\, a_{jj}$$
$$\hat{a}_{jk} = a_{jk} \quad j \neq k$$

where γ is some positive parameter. Equation (7.41) becomes

$$\hat{A}(x_0)(x_1 - x_0) = g \qquad (7.43)$$

For large γ, the matrix $\hat{A}$ will become diagonally dominant. As γ approaches zero, equation (7.43) will turn into the Newton-Raphson method. The Levenberg-Marquardt method has the basic feature of varying γ to select the optimal characteristics of the iteration. The basic Levenberg-Marquardt algorithm is summarized:

Levenberg-Marquardt Method

1. Set $k = 0$. Choose an initial guess x_0, γ and a factor α.

2. Solve the linear system of equation (7.43) to obtain x_{k+1}.

3. If $f(x_{k+1}) > f(x_k)$, reject x_{k+1} as the new approximation, replace γ by $\alpha\gamma$, and repeat step 2.

4. If $f(x_{k+1}) < f(x_k)$, accept x_{k+1} as the new approximation, replace γ by γ/α, set $k = k + 1$, and repeat step 2.

5. Terminate the iteration when

$$\|x_{k+1} - x_k\| < \varepsilon$$

In the problem of estimating a nonlinear waveform by a series of functions, the minimization function is given by:

$$\text{minimize} f = \sum_{i=1}^{N} \left[\sum_{k=1}^{m} \left[a_k e^{(b_k t_i)} \cos(\omega_k t_i + \theta_k) \right] - y_i \right]^2 \qquad (7.44)$$

where m is the number of desired modes of the approximating waveform, and $x = [a_1\, b_1\, \omega_1\, \theta_1\, \ldots a_m\, b_m\, \omega_m\, \theta_m]^T$. The results of a Prony fit can be used as initial guesses for the Levenberg-Marquardt by setting:

$$a_k = 2\,|B_k|$$
$$b_k = \text{real}(\lambda_k)$$
$$\omega_k = |\text{imag}(\lambda_k)|$$
$$\theta_k = \angle B_k$$

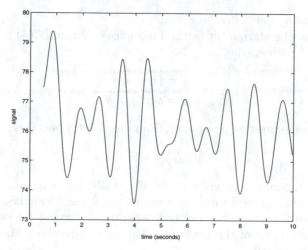

FIGURE 7.3
Waveform for Example 7.5

Example 7.5

For the waveform shown in Figure 7.3, apply the Prony method and the Levenberg-Marquardt method to estimate the eigenvalues of the system.

Solution 7.5

Due to the number of datapoints of the waveform in Figure 7.3, it is not possible to explicitly illustrate the exact steps required for this example. However, this example can still be used to highlight some of the characteristics of both the Prony and Levenberg-Marquardt methods. Applying the two methods yields the twenty estimated eigenvalues given below for the 6.0 second interval in Figure 7.4. The Prony method results were converted from the eigenvalues and residuals to the sinusoidal coefficients and used as the initial choices for the Levenberg-Marquardt iterations. Using the Prony results as a starting point, the Levenberg-Marquardt method required only three iterations to meet a tolerance of less than 0.1. The number of iterations may also vary based on choice of α and number of eigenvalues.

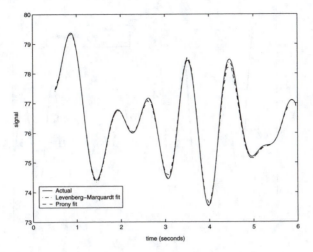

FIGURE 7.4
Simulated responses for Example 7.5: short interval

Levenberg-Marquardt	Prony
$-7.1216 \pm j28.4808$	$-3.1840 \pm j27.6972$
$-3.6418 \pm j21.9892$	$-2.0789 \pm j22.5951$
$-0.0443 \pm j14.6099$	$-0.2277 \pm j14.9743$
$-6.4881 \pm j16.1468$	$-6.4747 \pm j18.7613$
$-0.9911 \pm j11.3658$	$-1.0859 \pm j11.1334$
$0.1518 \pm j8.3250$	$0.1496 \pm j8.3281$
$-0.0811 \pm j7.0844$	$-0.0806 \pm j7.0873$
$0.1212 \pm j5.2737$	$0.1141 \pm j5.2718$
$-0.7399 \pm j1.0136$	$-0.8093 \pm j1.0119$
$-0.6668 \pm j3.2570$	$-0.7942 \pm j3.2916$

Note that the eigenvalues compare quite well between the two methods and the simulated responses correspond favorably with the actual value. There is a larger discrepancy in those modes that have a high oscillation frequency, but this has little impact on the simulated waveforms since these same modes are also highly damped as well and do not propagate over time.

The accuracy of the Prony method depends in large part on identifying the correct number of eigenvalues. In general, the larger the number of specified eigenvalues, the more accurate the results are. However, if too many eigenvalues are specified, the resulting set of eigenvalues has a tendency to contain "extraneous" eigenvalues, i.e., those eigenvalues that are highly damped and highly oscillatory. These eigenvalues are the result of over-fitting, or attempting to fit the nonlinear signal too closely to the sinusoidal function combination. They help the simulated waveform fit the actual waveform more closely, but do not really provide any additional usable data for control design or other applications.

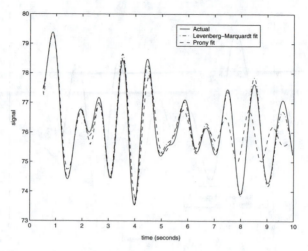

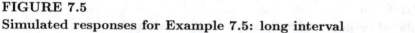

FIGURE 7.5
Simulated responses for Example 7.5: long interval

If the entire ten second waveform is used for approximating twenty eigenvalues, then a greater discrepancy appears in the eigenvalues and simulated response of the two methods. The time-domain simulation of the dynamic responses of these eigenvalue estimations is shown in Figure 7.5 and the eigenvalues are given below. The discrepancy between results can be a result of numerous factors. Firstly, the original waveform is the response of an inherently nonlinear system and both methods attempt to synthesize the waveform by a linear combination of sinusoidal functions. This becomes more complex over larger intervals where the nonlinearity becomes more dominant. The Levenberg-Marquardt is more accurate in this case because it attempts to estimate the coefficients of the sinusoidal combination through a nonlinear optimization, whereas the Prony method is a linear fitting scheme. However, since the Prony method is used to initialize the Levenberg-Marquardt method, it takes many more iterations to converge.

Levenberg-Marquardt	Prony
$-7.3056 \pm j12.0421$	$-2.2388 \pm j28.5914$
$-3.2844 \pm j11.7521$	$-1.8748 \pm j23.3888$
$-0.9222 \pm j10.2719$	$-0.7899 \pm j14.8773$
$-1.9858 \pm j9.6960$	$-4.0260 \pm j14.7060$
$-0.5478 \pm j8.6091$	$-1.5547 \pm j10.6161$
$-0.2854 \pm j7.7290$	$-0.1083 \pm j7.9489$
$0.0772 \pm j6.2749$	$-0.2509 \pm j6.6221$
$-0.1833 \pm j4.5367$	$-0.4762 \pm j4.2421$
$-0.2477 \pm j4.9520$	$-0.2863 \pm j5.2131$
-0.7625	-0.5351
-6.7867	-5.3497

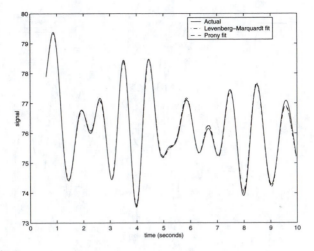

FIGURE 7.6
Simulated responses for Example 7.5: long interval with 28 eigen-values

The estimation was repeated, except that the number of specified eigen-values was raised from twenty to twenty-eight. In this case, both the Prony and Levenberg-Marquardt methods yielded time-domain results with compa-rable accuracy and similar eigenvalues. These results are shown in Figure 7.6. Once again, both methods generated a large number of highly oscillatory eigenvalues that are of questionable use.

Levenberg-Marquardt	Prony
$-1.4129 \pm j27.8578$	$-0.5210 \pm j28.1005$
$-1.8208 \pm j25.3685$	$-0.9840 \pm j25.6154$
$-1.0729 \pm j21.8998$	$-0.4650 \pm j22.0512$
$-2.6896 \pm j18.3350$	$-1.6967 \pm j18.6991$
$-0.8939 \pm j15.3934$	$-0.3244 \pm j15.5680$
$-1.0857 \pm j13.5562$	$-0.3214 \pm j13.6546$
$-0.5359 \pm j10.6563$	$-0.8611 \pm j10.9509$
$-0.2963 \pm j8.4528$	$-0.2922 \pm j8.4356$
$-0.2150 \pm j7.6795$	$-0.2122 \pm j7.6930$
$0.0149 \pm j6.2912$	$0.0151 \pm j6.2905$
$-0.0741 \pm j4.8703$	$-0.0835 \pm j4.8723$
$-0.9735 \pm j1.8668$	$-1.0158 \pm j1.8377$
$-0.5391 \pm j3.7206$	$-0.5784 \pm j3.7188$
-0.5758	-0.5751
-12.6099	-17.1626

If the first seven pairs of eigenvalues (imaginary part greater than 10) are neglected, the simulated waveform shown in Figure 7.7 is obtained. Even

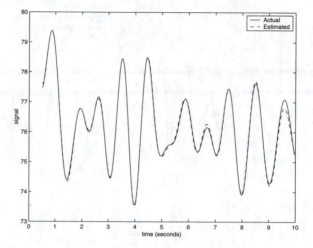

FIGURE 7.7
Simulated responses for Example 7.5: long interval with reduced eigenvalues

though the number of eigenvalues is greatly reduced, the resulting waveform is still quite accurate. This observation further supports the assertion that the highly oscillatory eigenvalues are superfluous and are not necessary for system analysis, but arise only as a numerical byproduct. ∎

The Levenberg-Marquardt method has been used widely in many applications that require the estimation of parameters for system identification purposes by minimizing the error between a set of measurements and a predicted system model. The extraction of modal information is only a subset of the rich field of applications for this method.

7.4 Power System Applications

7.4.1 Participation Factors

In the analysis of large scale power systems, it is sometimes desirable to have a measure of the impact that a particular state has on a selected system mode (or eigenvalue). In some cases, it is desirable to know whether a set of physical states has influence over an oscillatory mode such that control of that component may mitigate the oscillations. Another use is to identify which system components contribute to an unstable mode. One tool for identifying which states significantly participate in a selected mode is the method of *participation factors* [44]. In large scale power systems, participation factors can also be used to identify interarea oscillations versus those that persist only within localized regions (intra-area oscillations).

Participation factors provide a measure of the influence each dynamic state has on a given mode or eigenvalue. Consider a linear system

$$\dot{x} = Ax \qquad (7.45)$$

The participation factor p_{ki} is a sensitivity measure of the ith eigenvalue to the (k, k) diagonal entry of the system A matrix. This is defined as

$$p_{ki} \frac{\partial \lambda_i}{\partial a_{kk}} \qquad (7.46)$$

where λ_i is the ith eigenvalue and a_{kk} is the kth diagonal entry of A. The participation factor p_{ki} relates the kth state variable to the ith eigenvalue. An equivalent, but more common expression for the participation factor is also defined as

$$p_{ki} = \frac{w_{ki} v_{ik}}{w_i^T v_i} \qquad (7.47)$$

where w_{ki} and v_{ki} are the kth entries of the left and right eigenvectors associated with λ_i. As with eigenvectors, participation factors are frequently normalized to unity, such that

$$\sum_{k=1}^{n} p_{ki} = 1 \qquad (7.48)$$

When the participation factors are normalized, they provide a straightforward measure of the percent of impact each state has on a particular mode. Participation factors for complex eigenvalues (and eigenvectors) are defined in terms of magnitudes, rather than complex quantities. In the case of complex eigenvalues, the participation factors are defined

$$p_{ki} = \frac{|v_{ik}||w_{ki}|}{\sum_{i=1}^{n} |v_{ik}||w_{ki}|} \qquad (7.49)$$

In some applications, it may be preferred to retain the complex nature of the participation factors to yield both phase and magnitude information [21].

7.4.2   Modal Analysis

Power systems are inherently nonlinear, but in some instances may respond to well-tuned linear controls. In order to implement linear feedback control, the system designer must have a model of sufficiently reduced order from which to design the control. Several approaches to developing low order models have included dynamic equivalencing, eigenanalysis, and pole/zero cancellation. Frequently, however, the original system is too complex or the

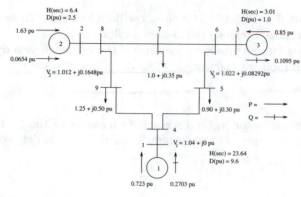

FIGURE 7.8
IEEE three-machine, nine-bus (WSCC) system

parameters are not known with enough accuracy to produce an adequate reduced order model. In this case, it is desirable to extract the modal information directly from the system response to a perturbation. Prony analysis has been used for estimating the modal content of power oscillations from measured ringdowns [4] [16] [31]. The choice of method used must consider the inclusion of nonlinearities, the size of the model that can be effectively utilized, and the reliability of the results.

Linearization eliminates the effects of the nonlinearities and is valid only for small perturbation around a stable operating point. On the other hand, methods that are applied directly to the nonlinear system simulation or field measurement results include the effects of nonlinearities. In full-state eigenvalue analysis, the size of the system model is currently limited to several hundred states with present computing capabilities. This means that a typical system containing several thousand buses and hundreds of machines must be reduced using dynamic equivalencing. Modal analysis techniques, such as the Prony and Levenberg-Marquardt methods, that operate directly on system output are not limited by system size. This means that standard time-domain-analysis results are directly usable. This eliminates the possibility of losing some of system modal content due to reduction.

Example 7.6

For the three-machine, nine-bus system shown in Figure 7.8, a solid three phase fault is applied on bus 8. The fault self-clears just prior to the critical clearing time and the system returns to the pre-fault configuration. The angular frequencies of the generators are shown in Figure 7.9. Calculate the modal content of this system using Prony analysis, the Levenberg-Marquardt method, and model linearization.

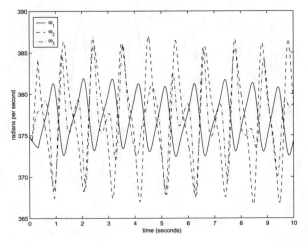

FIGURE 7.9
Short circuit angular frequencies

Solution 7.6 Since the fault is cleared just prior to the critical clearing time, the fault cannot truly be considered a "small-signal" perturbation. Since the generator 3 response appears to contain the most modal content, it will be used for the Prony and Levenberg-Marquardt methods. The state equations for this system are

$$\dot{\omega}_i = \frac{1}{M_i} \left(P_{m_i} - E_i^2 G_{ii} - E_i \sum_{j \neq i}^{n} E_j \left(B_{ij} \sin \delta_{ij} + G_{ij} \cos \delta_{ij} \right) \right) \quad (7.50)$$

$$\dot{\delta}_i = \omega_i - \omega_s \quad i = 1, \ldots, n \quad (7.51)$$

where $i = 1, 2, 3$ and the parameters are the same as Example 5.9. Since the post-fault system is the same as the pre-fault system, the eigenvalues may be calculated from the system Jacobian evaluated at the initial conditions. This system results in three oscillatory modes. The Levenberg-Marquardt and Prony methods are applied to ω_3 to obtain estimates for the eigenvalues as well. The Levenberg-Marquardt and Prony results are shown in Figure 7.10 and the eigenvalue estimates are given below.

Eigenvalues for Example 7.6		
Levenberg-Marquardt	Prony	Linearized
$-0.0514 \pm j16.3853$	$-1.6364 \pm j17.2116$	$0 \pm j12.3442$
$-0.0525 \pm j10.5390$	$-0.9521 \pm j11.2137$	$0 \pm j8.8129$
$0.0331 \pm j5.8934$	$-0.8415 \pm j6.2297$	$0 \pm j4.9124$

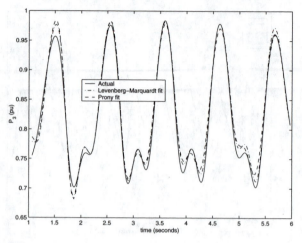

FIGURE 7.10
Simulated responses for Example 7.6

Although the Prony analysis waveform is far more damped than the actual simulation, it still captured the frequencies of the response reasonably accurately. The Levenberg-Marquardt method produced similar frequencies. The eigenvalues from the linearized system Jacobian are similar to both of the estimation methods. The Levenberg-Marquardt parameters are given below.

mode	A_i	B_i	C_i	D_i
1	1.4319	-0.0514	16.3853	1.2172
2	3.2779	0.0331	5.8934	0.8716
3	4.7951	-0.0525	10.5390	-1.3689

From the Levenberg-Marquardt parameters, it can be seen that the system response is impacted most significantly by modes 2 and 3, since these modes have the largest magnitude and are not significantly damped. Note also that the frequency of mode 3 is nearly twice that of mode 2, giving the system response an almost "third harmonic" look to it.

Since the impact of the fault on the system was severe, the responses contain significant nonlinearities and cannot be adequately modeled by linear approximations. This can be seen by the inability of both the Prony and the Levenberg-Marquardt methods in obtaining an exact reproduction of the system waveform. It may be possible to improve the fit by increasing the number of sinusoids (and therefore the number of modes) in the fitting function, but it is counter-productive to produce an estimated system with more states than the original system. ∎

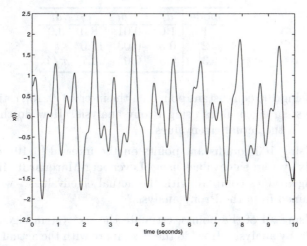

FIGURE 7.11
Waveform for Problem 3

7.5 Problems

1. Find the eigenvalues and eigenvectors of the following matrices

$$A_1 = \begin{bmatrix} 5 & 4 & 1 & 1 \\ 4 & 5 & 1 & 1 \\ 1 & 1 & 4 & 2 \\ 1 & 1 & 2 & 4 \end{bmatrix}$$

$$A_2 = \begin{bmatrix} 2 & 3 & 4 \\ 7 & -1 & 3 \\ 1 & -1 & 5 \end{bmatrix}$$

2. Find the eigenvalues of the follow matrix using the shifted-QR method.

$$A = \begin{bmatrix} 0 & 0 & 1 \\ 1 & 0 & 0 \\ 0 & 1 & 0 \end{bmatrix}$$

3. Generate the waveform shown in Figure 7.11 on the interval $t \in [0, 10]$ with a time step of 0.01 seconds

$$x(t) = \sum_{i=1}^{3} a_i e^{b_i t} \left(\cos c_i t + d_i \right)$$

where

mode	a_i	b_i	c_i	d_i
1	1.0	-0.01	8.0	0.0
2	0.6	-0.03	17.0	π
3	0.5	0.04	4.7	$\pi/4$

(a) Using 100 equidistant points on the interval $[0, 10]$, estimate the six system eigenvalues using Prony analysis. How do these compare with the actual eigenvalues?

(b) Using 100 equidistant points on the interval $[0, 10]$, estimate the six system eigenvalues using Levenberg-Marquardt. How do these eigenvalues compare with the actual eigenvalues? with those obtained from the Prony analysis?

(c) Using all of the points, estimate the six system eigenvalues using Prony analysis. How do these compare with the actual eigenvalues?

(d) Using all of the points, estimate the six system eigenvalues using Levenberg-Marquardt. How do these compare with the actual eigenvalues?

(e) Using all of the points, estimate the two dominant modes (two complex eigenvalue pairs) of the system response using Levenberg-Marquardt. Substitute the estimated parameters into

$$x(t) = \sum_{i=1}^{2} a_i e^{b_i t} \left(\cos c_i t + d_i \right)$$

and plot this response versus the three mode response. Discuss the differences and similarities.

References

[1] P. M. Anderson and A. A. Fouad, *Power System Control and Stability*. Ames, IA: Iowa State University Press, 1977.

[2] W. E. Arnoldi, "The principle of minimized iterations in the solution of the matrix eigenvalue problem," *Quart. Appl. Math.*, vol. 9, 1951.

[3] L. O. Chua and P. Lin, *Computer Aided Analysis of Electronic Circuits: Algorithms and Computational Techniques*. Englewood Cliffs, New Jersey: Prentice-Hall, Inc., 1975.

[4] R. Doraiswami and W. Liu, "Real-time estimation of the parameters of power system small signal oscillations," *IEEE Transactions on Power Systems*, vol. 8, no. 1, February 1993.

[5] H. W. Dommel and W. F. Tinney, "Optimal power flow solutions," *IEEE Transactions on Power Apparatus and Systems*, vol. 87, no. 10, pp. 1866-1874, October 1968.

[6] O. I. Elgerd, *Electric Energy System Theory, An Introduction*. New York, New York: McGraw-Hill Book Company, 1982.

[7] S. Eisenstate, M. Gursky, M. Schultz, and A. Sherman, "The Yale Sparse Matrix Package I: The symmetric codes," *International Journal of Numerical Methods Engineering*, vol. 18, 1982, pp. 1145-1151.

[8] J. Francis, "The QR Transformation: A unitary analogue to the LR Transformation," *Comp. Journal*, vol. 4, 1961.

[9] C. W. Gear, *Numerical Initial Value Problems in Ordinary Differential Equations*, Englewood Cliffs, New Jersey: Prentice-Hall, Inc., 1971.

[10] C. W. Gear and L. R. Petzold, "ODE methods for the solution of differential/algebraic systems," *SIAM Journal of Numerical Analysis*, vol. 21, no. 4, pp. 716-728, August 1984.

[11] A. George and J. Liu, "A fast implementation of the minimum degree algorithm using quotient graphs," *ACM Transactions on Mathematical Software*, vol. 6, no. 3, September 1980, pp. 337-358.

[12] A. George and J. Liu, "The evoluation of the minimum degree ordering algorithm," *SIAM Review*, vol. 31, March 1989, pp. 1-19.

[13] H. Glavitsch and R. Bacher, "Optimal power flow algorithms," *Control and Dynamic Systems*, vol. 41, part 1, *Analysis and Control System*

Techniques for Electric Power Systems, New York: Academic Press, 1991.

[14] G. H. Golub and C. F. Van Loan, *Matrix Computations*, Baltimore: Johns Hopkins University Press, 1983.

[15] G. K. Gupta, C. W. Gear, and B. Leimkuhler, "Implementing linear multistep formulas for solving DAEs," Report no. UIUCDCS-R-85-1205, University of Illinois, Urbana, Illinois, April 1985.

[16] J. F. Hauer, C. J. Demeure, and L. L. Scharf, "Initial results in Prony analysis of power system response signals," *IEEE Transactions on Power Systems*, vol. 5, no. 1, February 1990.

[17] M. Ilic and J. Zaborszky, *Dynamics and Control of Large Electric Power Systems*, New York: Wiley-Interscience, 2000.

[18] D. Kahaner, C. Moler, and S. Nash, *Numerical Methods and Software*, Englewood Cliffs, NJ: Prentice-Hall, 1989.

[19] R. Kress, *Numerical Analysis*, New York: Springer-Verlag, 1998.

[20] V. N. Kublanovskaya, "On some algorithms for the solution of the complete eigenvalue problem," *USSR Comp. Math. Phys.*, vol. 3, pp. 637-657, 1961.

[21] P. Kundur, *Power System Stability and Control*. New York: McGraw-Hill, 1994.

[22] J. Liu, "Modification of the minimum-degree algorithm by multiple elimination," *ACM Transactions on Mathematical Software*, vol. 11, no. 2, June 1985, pp. 141-153.

[23] H. Markowitz, "The elimination form of the inverse and its application to linear programming," *Management Science*, vol. 3, 1957, pp. 255-269.

[24] A. Monticelli, "Fast decoupled load flow: Hypothesis, derivations, and testing," *IEEE Transactions on Power Systems*, vol. 5, no. 4, pp 1425-1431, 1990.

[25] J. Nanda, P. Bijwe, J. Henry, and V. Raju, "General purpose fast decoupled power flow," *IEEE Proceedings-C*, vol. 139, no. 2, March 1992.

[26] J. M. Ortega and W. C. Rheinboldt, *Iterative Solution of Nonlinear Equations in Several Variables*, San Diego: Academic Press, Inc., 1970.

[27] A. F. Peterson, S. L. Ray, and R. Mittra, *Computational Methods for Electromagnetics*, New York: IEEE Press, 1997.

[28] M. J. Quinn, *Designing Efficient Algorithms for Parallel Computers*, New York: McGraw-Hill Book Company, 1987.

[29] O. R. Saavedra, A. Garcia, and A. Monticelli, "The representation of shunt elements in fast decoupled power flows," *IEEE Transactions on Power Systems*, vol. 9, no. 3, August 1994.

[30] P. W. Sauer and M. A. Pai, *Power System Dynamics and Stability*, Upper Saddle River, New Jersey: Prentice-Hall, 1998.

[31] J. Smith, F. Fatehi, S. Woods, J. Hauer, and D. Trudnowski, "Transfer function identification in power system applications," *IEEE Transactions on Power Systems*, vol. 8, no. 3, August 1993.

[32] G. Soderlind, "DASP3–A program for the numerical integration of partitioned stiff ODEs and differential/algebraic systems," Report TRITA-NA-8008, The Royal Institute of Technology, Stockholm, Sweden, 1980.

[33] D. C. Sorensen, "Implicitly restarted Arnoldi/Lanzcos methods for large scale eigenvalue calculations," in D. E. Keyes, A. Sameh, and V. Venkatakrishnan, editors, *Parallel Numerical Algorithms: Proceedings of an ICASE/LaRC Workshop, May 23-25, 1994, Hampton, VA*, Kluwer, 1995.

[34] D. C. Sorensen, "Implicit application of polynomial filters in a k-step Arnoldi method," *SIAM J. Mat. Anal. Appl.*, vol. 13, no. 1, 1992.

[35] B. Stott and O. Alsac, "Fast decoupled load flow," *IEEE Transactions on Power Apparatus and Systems*, vol. 93, pp. 859-869, 1974.

[36] G. Strang, *Linear Algebra and Its Applications*, San Diego: Harcourt Brace Javanonich, 1988.

[37] W. Tinney and J. Walker, "Direct solutions of sparse network equations by optimally ordered triangular factorizations," *Proceedings of the IEEE*, vol. 55, no. 11, November 1967, pp. 1801-1809.

[38] L. Wang and A. Semlyen, "Application of sparse eigenvalue techniques to the small signal stability analysis of large power systems," *IEEE Transactions on Power Systems*, vol. 5, no. 2, May 1990.

[39] D. S. Watkins, *Fundamentals of Matrix Computations*. New York: John Wiley and Sons, 1991.

[40] J. H. Wilkinson, *The Algebraic Eigenvalue Problem*. Oxford, England: Clarendon Press, 1965.

[41] M. Yannakakis, "Computing the minimum fill-in is NP-complete," *SIAM Journal of Algebraic Discrete Methods*, vol. 2, 1981, pp. 77-79.

[42] T. Van Cutsem and C. Vournas, *Voltage Stability of Electric Power Systems*, Boston: Kluwer Academic Publishers, 1998.

[43] R. S. Varga, *Matrix Iterative Analysis*. Englewood Cliffs, New Jersey: Prentice-Hall, Inc., 1962.

[44] G. C. Verghese, I. J. Perez-Arriaga, and F. C. Schweppe, "Selective modal analysis with applications to electric power systems," *IEEE Transactions on Power Systems*, vol. 101, pp. 3117-3134, Sept. 1982.

Index

Adam's methods, 126
  Bashforth, 127
  Moulton, 128
admittance matrix, 151
  reduced, 152
    fault-on, 156
    post-fault, 156
    pre-fault, 156
Arnoldi methods, 213–220
  k-step, 214
  $k + p$-step, 214
  implicitly restarted Arnoldi, 213
  modified Arnoldi, 213

back substitution, 7
backward Euler, 122
bad data, 168, 174
  chi-squared test, 175
Berry, 104

complete pivoting, 20, 21
condition number, 22
conjugate gradient, 28–33
  preconditioning, 32–33
continuation power flow, 69
convergence
  fixed point iteration, 40
  Newton-Raphson, 45
  QR method, 213
  relaxation methods, 24
Cramers rule, 3
Crout's algorithm, 11

decoupled power flow, 64
descriptor systems, 148
difference methods, 147
differential-algebraic systems, 148–149

direct methods, 3
dishonest Newton, 51
domain of attraction, 41

eigenanalysis, 205
eigenvalue, 205
eigenvector, 205
elementary row operations, 4
equal incremental cost, 185

fixed point iteration, 38
forward elimination, 5
forward Euler method, 119

Gauss' method, 20
Gauss-Seidel method, 23
Gaussian distribution, 172
Gaussian elimination, 4–9
  back substitution, 7
  forward elimination, 5
Gear's methods, 129–130, 143

Hessenberg, 206
Householder matrix, 206

idempotent matrix, 175
initial value problem, 117
integration step size selection, 147
interarea oscillations, 231
iterative methods, 4, 37

Jacobi method, 23

Krylov subspace, 214

Lagrange multipliers, 178
least squares, 168–178, 223
  weighted, 171–174

Levenberg-Marquardt method, 223–230

local truncation error, 130–134
    backward Euler, 133
    forward Euler, 132
    multistep methods, 134
    trapezoidal, 133, 148

LU factorization, 9–22, 53, 89, 157, 169, 178, 206
    backward substitution, 12
    complete pivoting, 20, 21
    Crouts algorithm, 11
    forward substitution, 12
    Gauss' method, 20
    partial pivoting, 16
    permutation matrix, 17
    sparse, 89

Markowitz, 98
mid-term stability analysis, 159–161
minimum degree, 98
modal analysis, 231
model identification, 220
multistep methods, 120–130
    local truncation error, 134
    numerical stability, 138

Newton-Raphson, 157, 198, 224
Newton-Raphson method, 42
    dishonest, 51
nonlinear least squares, 177–178, 223
numerical integration, 117–150
    absolute stability, 138
    Adam's methods, 126
        Bashforth, 127
        Moulton, 128
    backward Euler, 122
    consistency, 127
    exactness constraints, 128
    explicit method, 122
    forward Euler, 119, 121
    Gear's methods, 129–130
    implicit method, 122, 125
    local truncation error, 130–134
    multistep methods, 120–130

numerical accuracy, 117
one step methods, 118
order, 119
predictor-corrector, 129
Runge-Kutta, 119, 124
stability, 134–143
step size selection, 145–148
stiff systems, 141–143
Taylor series methods, 118–120
time step, 117
trapezoidal method, 122
numerical stability, 134–143
    backward Euler, 136
    forward Euler, 135
    multistep methods, 138
    relative stability, 142

ODEs, 117
optimal power flow, 184–195

parallel processing, 23
partial pivoting, 16
participation factors, 230–231
permutation matrix, 9, 17
point of attraction, 40
power flow, 52, 151
    continuation method, 69
    decoupled, 64
    fast decoupled, 66
    flat start, 57
    Newton-Raphson, 53
    optimal, 184–195
    power flow equations, 52
    regulating transformers, 60
    three-phase, 76
Prony method, 221–223, 225–230
PV curves, 69

QR method, 206–213, 216
    convergence, 213
    deflation, 213
    shifted, 212

rate of convergence, 46
regulating transformers, 60

relaxation methods, 23–28
    convergence, 24
    Gauss-Seidel, 23
    Jacobi, 23
Runge-Kutta, 119

SCADA, 195
signal to noise ratio, 223
similarity transformation, 209
sparse matrix, 81–111
    fills, 89
    linked list, 82
    matrix representation, 87–89
    ordering schemes, 89–107
    Scheme 0, 97
    Scheme I, 98–104
    Scheme II, 104–106
    sparse storage, 81–87
state estimation, 178, 195–199
    observability, 195
steepest descent, 178–183
structure-preserving model, 159
successive overrelaxation, 28

Taylor series methods, 118–120
three-phase power flow, 76
Tinney I, 98
Tinney II, 104
transient stability analysis, 150–157
trapezoidal method, 122, 156

unitary matrix, 206

weighted least squares, 171–174